INTERNATIONAL FORUM SERIES

Volume 1

Traffic, transportation and urban planning

Edited versions of papers originally presented at
an international conference in Tel Aviv of
the Institute of Transportation Engineers,
the Association of Engineers and Architects and
the International Technical Cooperation Centre

George Godwin

© Conference Associates, 1981

First published 1981 by
George Godwin Limited
The book publishing subsidiary
of The Builder Group
1–3 Pemberton Row
Red Lion Court
Fleet Street
London EC4P 4HL

British Library Cataloguing in Publication Data

Traffic, transportation and urban planning—
  (International forum series)
  Vol. 1
  1. City planning—Congresses
  I. Hakkert, A. S.   II. Series
  711′.4 HT166

ISBN 0–7114–5713–1

Printed and bound in Great Britain at
The Pitman Press, Bath

VOLUME ONE

Contents

PART III: REVIEW PAPERS

Introduction and acknowledgments

The forty-six papers in Volumes 1 and 2 of "Traffic, transportation and urban planning" derive from an international conference held in Tel Aviv on 17-22 December 1978. The conference theme was "The integration of traffic and transportation engineering in urban planning", and its joint sponsors were the Institute of Transportation Engineers (ITE), Israel section, the Association of Engineers and Architects in Israel, and the International Technical Cooperation Centre (ITCC). A total of seventy-one papers were submitted for the conference. Many of these were subsequently revised and the editors responsible for the selection in this printed record were Dr A.S. Hakkert, P. Ben Shaul, Dr Dan Link and Professor S. Reichmann.

The conference provided a forum for planners, urbanists and transportation engineers and dealt with ways by which the urban environment can be improved through transportation. The four hundred participants from twenty countries gave proof that worldwide efforts through interdisciplinary cooperation between all parties concerned are of the greatest importance in the elimination or at least the reduction of conflicts between traffic and environment. The conference concluded that only by the integration of all the disciplines involved can the quality of the environment be improved.

All those who participated in arranging the programme, the international advisory committee and the local organising committee deserve recognition for their assistance. Thanks are also due to the authors from fifteen countries who submitted material for this printed record. Special thanks for their invaluable contributions are due to the president and the executive director of the International Institute of Transportation Engineers, William Moreau and Thomas Brahms; the three guest lecturers, Professor Harold L. Michael (United States), Professor Nathaniel Lichfield (United Kingdom) and Professor Ernest Fiala (West Germany); the five general reporters, Professor H.G. Retzko (West Germany), Professor Paul H. Bendsten (Denmark), Sam Cass (Canada), B. Beukers (the Netherlands) and Professor John E. Baerwald (United States); and the four editors whose names are listed above.

A resolution was adopted to the effect that further international and regional meetings should continue to deal with specific themes of the subject of this conference. It is hoped that the papers presented in this printed record will provide the impetus for that.

Max G. Shifron
General Chairman

The contributors

□ Professor David C. Andreassend, author of Paper 39, is Associate Professor of Transportation Engineering, Asian Institute of Technology, Bangkok, Thailand.

□ Osnat Arnon, author of Paper 43, is a member of the Environmental Protection Service, Ministry of the Interior, Israel.

□ H. Ayad, co-author of Paper 9, is a member of the Department of Civil Engineering, University of Vermont, Burlington, Vermont, USA.

□ Professor John E. Baerwald, author of Paper 24, is Professor of Transportation and Traffic Engineering, University of Illinois at Urbana-Champain, USA.

□ Professor J.J. Bakker, author of Paper 31, is Head of the Department of Civil Engineering, University of Alberta, Canada.

□ Hans B. Barbe, author of Paper 3, is President of Hans B. Barbe and Associates, Zürich and Winchester.

□ J. Bar-Ziv, co-author of Paper 38, is a member of the Road Safety Centre, Technion, Haifa, Israel.

□ Dr Margaret C. Bell, author of Paper 30, is a member of the Transport Operations Research Group, Department of Civil Engineering, University of Newcastle-upon-Tyne, England.

□ Moshe Ben-Akiva, co-author of Paper 10, is a member of the Transportation Research Institute, Technion, Haifa, Israel.

□ Professor P.H. Bendtsen, author of Paper 21, is from the Technical University of Denmark.

□ C.R. Berger, co-author of Paper 28, is on the staff of Sperry System Management, Great Neck, N.Y., USA.

□ Eric L. Bers, author of Paper 11, is General Engineer of the Interstate Commerce Commission, Washington, D.C., USA.

□ B. Beukers, author of Paper 23, is Director of the Traffic and Transportation Engineering Division, Rijkswaterstaat, Ministry of Transport and Public Works, the Netherlands.

□ J.A. Bourdrez, co-author of Paper 29, is a member of the Netherlands Economic Institute in Rotterdam.

□ Dr John P. Braaksma, author of Paper 46, is an Associate Professor in the Department of Civil Engineering, Carleton University, Ottawa, Canada.

□ Valerie Brachya, co-author of Paper 43, is a member of the Environmental Protection Service, Ministry of the Interior, Israel.

☐ R.E. Brindle, author of Paper 40, is on the staff of the Australian Road Research Board, Melbourne, Australia.

☐ Richard J. Brown, author of Papers 4 and 12, is a Senior Lecturer in the Department of Civil Engineering, University of the Witwatersrand, Johannesburg, South Africa.

☐ Friedemann Brühl assisted Professor Retzko in the preparation of Paper 20.

☐ K.J. Button, co-author of Paper 15, is a member of the Department of Economics, Loughborough University, England.

☐ Sam Cass, author of Paper 22, is Commissioner for Roads and Traffic for Metropolitan Toronto, Canada.

☐ António José de Castilho, author of Paper 8, is Head of the Traffic and Road Safety Division, Laboratório Nacional de Engenharia Civil, Lisbon, Portugal.

☐ Erella Daor, author of Paper 13, is a research associate in the Transportation Research Institute, Technion, Haifa, Israel.

☐ Joseph Efrat, author of Paper 35, is an architect and town planner in Herzliya, Israel.

☐ Professor Ernst Fiala, author of Paper 34, is Director (Research and Development) of Volkswagenwerk, West Germany.

☐ L.R. Ford, co-author of Paper 28, is on the staff of the General Research Corporation, Santa Barbara, California, USA.

☐ Reinhard Forst assisted Professor Retzko in the preparation of Paper 20.

☐ Nathan H. Gartner, author of Paper 25, is a member of the Federal Highway Administration, Washington, DC, USA.

☐ B.B. Gragg, co-author of Paper 28, is on the staff of Sperry Systems Management, Great Neck, N.Y., USA.

☐ Dr A.S. Hakkert, co-author of Paper 38, is a staff member of the Road Safety Centre, Technion, Haifa, Israel.

☐ L.W. Hambly, co-author of Paper 28, is on the staff of the General Research Corporation, Santa Barbara, California, USA.

☐ W.E. Johnson, co-author of Paper 28, is on the staff of Sperry Systems Management, Great Neck, N.Y., USA.

☐ Shogo Kawakami, author of Paper 17, is a member of the Department of Civil Engineering, Nagoya University, Nagoya, Japan.

☐ Harmut Keller, co-author of Paper 45, is from the Technical University, Munich, West Germany.

☐ L.H. Klassen, co-author of Paper 29, is a member of the Netherlands Economic Institute in Rotterdam.

☐ Fumihiko Kobayashi, author of Paper 27, is from the Toyota Motor Company Ltd, Toyota, Japan.

☐ Katsunao Kondo, co-author of Paper 7, is a member of the Department of Civil Engineering, Fukuyama University, Fukuyama, Japan.

☐ Professor Karl Krell, author of Paper 5, is Director of the Highway Research Institute (BAST) Cologne, West Germany.

☐ Professor Rudolf Lapierre, author of Paper 26, is an engineer on the staff of the Federal Ministry of Transport, Bonn, West Germany.

☐ L. R. Leembruggen, author of Paper 32, is Managing Director of the firm Elroy Engineering in Australia.

□ Herbert S. Levinson, author of Paper 2, is an American transportation engineer.

□ Professor Nathaniel Lichfield, author of Paper 1, is the Senior Partner of Nathaniel Lichfield and Partners, Hampstead, London, England.

□ Dr Dan Link, author of Paper 33, is Chief Engineer (Transportation Projects) of the Ministry of Transport, Jerusalem, Israel.

□ Marvin L. Manheim, co-author of Paper 10, is on the staff of Massachusetts Institute of Technology, Cambridge, Massachusetts, USA.

□ Harold Marks, author of Paper 44, is Director of Traffic and Transportation, Gruen Associates, Los Angeles, California, USA.

□ Professor H.L. Michael, author of Paper 37, is from Purdue University, USA.

□ Nissim Moses, co-author of Paper 43, is an accoustics engineer on the staff of Israel Aircraft Industries.

□ Wilfried Müller, co-author of Paper 45, is from the Technical University, Munich, West Germany.

□ J.C. Oppenlander, co-author of Paper 9, is a member of the Department of Civil Engineering, University of Vermont, Burlington, Vermont, USA.

□ Professor Alessandro Orlandi, author of Paper 18, is Permanent Professor, Department of Transportation, Faculty of Engineering, University of Bologna, Italy.

□ G.R. Otten, co-author of Paper 29, is a member of the Netherlands Economic Institute in Rotterdam.

□ A.D. Pearman, co-author of Paper 15, is on the staff of the School of Economic Studies, University of Leeds, England.

□ Michael C. Poulton, author of Paper 41, is an Assistant Professor in the School of Community and Regional Planning, University of British Columbia, Canada.

□ R.A. Reiss, co-author of Paper 28, is on the staff of Sperry Systems Management, Great Neck, N.Y., USA.

□ Professor Hans-Georg Retzko, author of Paper 20, is from the Technical University, Darmstadt, West Germany.

□ Professor E.A. Rose, co-author of Paper 6, is Head of the Departmant of Architectural, Planning and Urban Studies, University of Aston, Birmingham, England.

□ Nils T.I. Rosén, author of Paper 36, is a civil engineer on the staff of Traffic Planning Ltd, Saltsjöbaden, Sweden.

□ Ilan Salomon, co-author of Paper 10, is on the staff of Massachusetts Institute of Technology, Cambridge, Massachusetts, USA.

□ Tsuna Sasaki, co-author of Paper 7, is a member of the Department of Transportation Engineering, Kyoto University, Kyoto, Japan.

□ Keith T. Solomon, author of Paper 42, is Senior Lecturer in Highway and Traffic Engineering, Caulfield Institute of Technology, Melbourne, Australia.

□ Professor Peter R. Stopher, co-author of Paper 19, is a faculty member of Northwestern University, USA.

□ P. Truelove, co-author of Paper 6, is a member of the Department of Architectural, Planning and Urban Studies, University of Aston, Birmingham, England.

□ J.A. van der Vlist, co-author of Paper 29, is a member of the Netherlands Economic Institute in Rotterdam.

□ Dr Velibor Vidakovic, author of Paper 14, is on the staff of Delft University of Technology and a member of the Department of Public Works, Amsterdam, the Netherlands.

☐ Chester G. Wilmot, co-author of Paper 19, is on the staff of the National Institute of Transport and Road Research, Council for Scientific and Industrial Research, Pretoria, South Africa.

☐ Yacov Zahavi, author of Paper 16, is a consultant based in Washington, DC, USA.

☐ Ora Zeitlin, co-author of Paper 43, is an acoustics engineer on the staff of Israel Aircraft Industries.

PART I: INTEGRATING TRANSPORTATION INTO
THE OVERALL PLAN

Transportation and land-use planning

1

Nathaniel Lichfield   United Kingdom

SYNOPSIS

The main weight of the contributions to this volume is from traffic and transportation engineers and planners who are concerned with the movement of people and goods in urban areas. In these contributions there is an awareness of the need to consider such movement of people and goods in the wider context of the urban system, with due regard to such matters as the environment, conservation of resources and quality of life. Put another way, the contributors recognise that traffic and transportation engineering and planning must be integrated with the wider concerns of urban planning.

In this paper the recognition is the same: the integration of planning for movement with the wider land-use planning. But the telescope is reversed and the view is from the other end, that of land-use planning.

To develop this theme, it is necessary first to comprehend, in contemporary terms, the nature of land-use planning and the role of transportation within it. But since the world does not practise land-use and transportation planning uniformly in these contemporary terms, and countries are at various stages of development in the process, it is useful to show how traffic and transportation planning has evolved in its relationships with land-use planning.

From this general review we arrive at three particular aspects in which the integration can be more fully demonstrated: in the contribution land-use planning can make to minimising traffic problems; the place of traffic in the growing field of urban impact analysis; and the place of traffic in the growing field of comprehensive evaluation and appraisal of traffic and transportation projects.

While the integration in the terms so described is of general relevance, there are particular lessons to be learned for developing countries, and the lessons are set out in the conclusion.

NATURE OF LAND-USE PLANNING

While the theme of the volume relates to "urban" planning, the title for this paper is the term more familiar in transportation planning, namely, "land use". But many other synonyms could have been used with equal justification, such as town and country, urban and regional, spatial, physical, environmental, development or human settlement planning. These terms do convey some difference in emphasis, however. For example, "town and country" or "land use" planning conveys a preoccupation with the allocation of uses on the earth's surface; "urban and regional" planning more popularly denotes socio-economic activities associated with land use; "physical" stresses that the impact of such planning is on the physical expression of activities in the built environment; "environmental" stresses the concern with matters affecting the human being.

But while for our purpose the terms can be used interchangeably, it is relevant to add that the terminology itself denotes how land-use planning has evolved from its origins around the turn of the century. Evolution within various countries has not been uniform, so one country, Britain, will be used as an illustration, particularly as in this country the transformation over this century has been dramatic. It started before the first world war with a concern for "town" planning which in fact related only to the peripheral expansion of towns and not to the established core. Thereafter in the 1930s the whole of the country's surface was taken into account in town and country planning. This was followed by the transformation from negative to positive planning under the title "development" planning after the second world war.

Then, after some twenty years, followed the attempt to enrich the social and economic aspects of such "development" planning alongside physical planning, together with the recognition that such planning was to be seen both as part of the corporate planning of municipalities and the urban management or government of communities.[1]

But even this contemporary concept is in many respects not as advanced as that adopted internationally in 1976 in the United Nations Vancouver Conference on the habitat. Here there was a near unanimous declaration of principles by 132 countries, which included recommendations that governments and international organisations make every effort to take action on certain guide lines. Among this was the following statement: "It is the responsibility of government to prepare spatial strategy plans and adopt human settlement policies to guide the socio-economic development efforts. Such policies must be an essential component of an overall development strategy, linking and harmonising them with policies on industrialisation, agriculture, social welfare and environmental and cultural preservation, so that each supports the other in the progressing improvement of the well-being of all mankind. A human settlement policy must seek harmonious integration or coordination of a wide variety of components including, for example, population growth or distribution, employment, shelter, land use, infrastructure, and services."[2] This declaration of principles was supported by some 80 pages of text conveying detailed recommendations with specific indications for their implementation.

This brief review of the evolving role and nature of land-use planning, with the indication that in any particular country the state of the art will vary within this spectrum, relates to what could be called the theory and principles of land use and other forms of planning. But in any particular country this is not the important dimension in practice. Rather it is the degree to which the laws of the country have authorised the practice of such planning by central and local government (for without such authority the principles are of little avail); and the degree to which in practice such plans are adopted and implemented (and there is a general weakness in implementation around the world).

Returning to the focus of this paper, there can thus be no universally applicable demonstration of how traffic and transportation planning can be incorporated into land-use planning. Accordingly, for our purpose we must keep to the principles of advanced contemporary practice. This will be attempted in the following section which presents a concept of land-use planning within which can be seen the role of traffic and transportation planning.

ROLE OF TRANSPORTATION WITHIN LAND-USE PLANNING

The traffic and transportation engineer and planner sees the movement of goods and people as one of the urban and regional sub-systems, of which employment and shopping are others. But perhaps even this is overstating the situation, for it is comparatively rare that he sees the whole of movement as one sub-system. Indeed, there has been the tendency to see fragments in isolation within the sub-system, such as subdivisions of road traffic (the private motor vehicle, public transport, cycles and pedestrians) and, quite separate from these, rail traffic (both long-haul and commuter), airports, ports, shipping etc.

But to the land-use planner, planning of the total system is the aim, within which integrated traffic and transportation is a part, albeit an important part. There are many ways of visualising the total regional system and the following is but one of the possible typologies.[3]

First comes the physical system in which can be noted seven distinct elements: the land on which all human settlement is founded; the minerals beneath the land which are the prime source of fossil energy; other natural resources which are not only necessary for life (sun, air, wind) but also energy sources; the man-made buildings and the surrounding places for living, work and recreation; the infrastructure for servicing those buildings and places (water, sewerage, gas, electricity, etc.); means of transportation to convey people from various fixed points to other (home to work, home to school, etc.); and communication along wires or through the air, which enables people to have transactions with each other without having to travel, and is therefore a substitute for it.

This physical system is the substructure for the real purpose of human settlement: people's activities within it. Associated with each of these physical elements are corresponding social, economic and cultural activities, such as production in factories, consumption in restaurants, education in schools, etc. More relevant to our purpose, there is the movement of people and goods by the various transportation modes and their substitute, the non-face-to-face linkages between people carrying out their various activities by use of communications rather than transportation.[4]

Seen this way, transportation is the active movement of people and goods along the various centres or zones of socio-economic activity. This is the essence of the land-use/transportation systems and the nub of the land-use/transportation studies. This relationship can be described precisely through modelling the relationship at any moment between the potential for traffic generation by a land use and the capability of the various transportation modes to carry that traffic.[5] The traffic generated is thus a joint consequence of the land-use potential and the capability of the transport system to carry the traffic, both in terms of quality, categories, desired origins, destinations, desired times of travel, etc. And the introduction of a transport capability will in itself generate the possibility of socio-economic activities on land; and the generation of such activities will stimulate the provision of transport capabilities.

But this comprehension of the total urban and regional system, and the role of transportation within it, is only a starting point to the purpose for which studies are made: the planning of the urban and regional system as a whole. Essentially, in such planning, we ask how the system is likely to evolve, given no planning intervention; and, if this is considered undesirable in terms of the objectives of the community, we aim to devise a more desirable system. To do this, we sketch alternative futures for the system, test them for feasibility, evaluate them in order to choose one which is preferred, and proceed to implementation. Following implementation, we monitor in order to review the proposals in the light of the experience gained since the plans were prepared.

For the purpose of making, implementing and reviewing plans on these lines it is necessary to carry out some formal planning processes. There are many ways of doing this, none of which is standard. But there are none the less generally accepted models of the rational process for urban and regional planning, of which one typical example is as follows.[6]

1. The planning content
2. Preliminary recognition and definition of problems
3. Decision to act and definition of planning task
4. Data collection, analysis and forecasting
5. Identification of problems, opportunities and objectives
6. Identification of constraints
7. Formulation of operational criteria for design
8. Generation of alternatives
9. Tests on alternatives
10. Plan evaluation as an aid to choice
11. Decision making
12. Work up preferred plan
13. Implementation programme
14. Plan implementation
15. Monitoring and review of planned developments through time
16. Politics
17. Consultation
18. Public participation

It would be of interest here to note the role within such a model of the conventional land-use/transportation studies. In some instances they have been entirely detached, and in others there have been attempts at integration.[7] But what is of greater interest is the evolution. Initially, transportation planners would take as inputs in their modelling the proposed land use and then derive the traffic flows which would be generated, thus tending to overlook the interaction between the land-use potential and transportation capability mentioned above. This has been recognised in later studies in which attempts have been made to take the reverse relationship, that of transport on land use, into account as well. More specifically, the aim is that both transport and land-use strategies and policies should be evolved together, and their mutual interaction borne in mind.[8]

An example of this change in emphasis is seen in the series of studies examining the impact of the San Francisco Bay rapid transit system (BART) opened in 1972.[9] One such study concluded that BART has had a considerable effect on the spatial arrangement of people and activities within the San Francisco Bay area despite insufficient regard on the part of the planning authorities for the importance of the relationship between transport and land-use planning, by omitting to encourage desired changes through appropriate land-use zoning policies.[10]

Despite these changes of emphasis, however, the modelling of land-use impacts of transport schemes and accessibility changes is still in its infancy, partly because of the inherent difficulties in identifying activities especially sensitive to the transport changes in prospect, and the quantitative effect of given travel cost charges on the location of firms and scale of output decisions.

As to prediction, the best that usually can be done is to measure prospective changes in the accessibility of zones to areas of destination and sources of supply, and make a judgment of likely shifts in locations of households and firms with respect to those zones which undergo the greatest change in accessibility (improved or reduced).[11]

Moving on from these technical minutiae, a contrast needs to be drawn between the role of land-use and transportation planning within this planning process. In transportation planning there is often a direct link in terms of agencies and skills between those concerned with the planning and those concerned with the implementation of the projects which are the end product of the planning, e.g. the roads and railways. In this way the planning is seen clearly as a lead into construction. But in the general practice of urban and regional planning there is a sharper distinction between the plan makers and the plan implementers. Development and construction on the ground is the role of development agencies and their professionals (shopping, industry, roads, etc), and not of the planning agencies. To be sure, there are instances where an urban and regional planning authority is also the implementing authority, e.g. in the provision of housing or a shopping centre in a town; and there is full integration when a new town development corporation builds a new town.[12] But since this is the exception rather than the rule, there is in practice a division of function between those who develop and construct, and those who do the preliminary planning. For these reasons, the realities of implementation in urban and regional planning are less strong than in transportation and traffic planning. This is a hurdle in integration.

EVOLUTION OF TRAFFIC AND TRANSPORTATION PLANNING
IN LAND-USE PLANNING

So far we have described the role of transportation planning in land-use planning in a somewhat "idealised" way, simply because it is not possible to generalise either for countries or for a period of time. But having done so, it is now helpful to consider how the role has evolved over the past quarter of a century or so, for this is a way of emphasising important considerations in the integration of transportation and land-use planning: it was because they were important that the integration took place. We do so first by considering changes in transportation planning itself, and then changes in the economic environment which have affected both transportation and land-use planning. For the reasons given above, in relation to land use planning, it is simpler to consider the evolution in only one country. Here again we take Britain, but the story is not atypical.

Initially, the emphasis was naturally on planning for growth in new modes of transport to supplement the traditional and established forms. Thus, in the early part of the century there was the need to introduce highways to cater for the growing motor traffic, alongside the former modes of walking, cycling, canals, rail, ports, etc. After the second world war there arose the need to plan for airports and the associated surface access facilities, with the extraordinary increase of aircraft movements. During the early days, which continued well after the second world war, the approach to planning for the motor vehicle was a matter of trying to predict the demand and cater for it.[13] But in the 1940s it was also recognised that vehicle trips had various roles within the total motor vehicle traffic system, so that segregation in a road hierarchy was needed. An early formulation was arterial, sub-arterial and local.[14]

In the 1960s it was generally recognised that, while segregation would continue to be needed, it would not be practicable to cater for road traffic in full. The limitations were of various kinds. If roads in built-up areas were allowed to carry traffic to their full capacity there would be adverse influences on the local environment in terms of noise, atmospheric pollution, vibration, etc. This led to the concept of "environmental" as opposed to "crude" capacity for such roads, with a modified hierarchy of distribution and access roads.[15] But even given the environmental constraints, the urban surgery needed to cater fully for the motorist in towns beyond a certain size created financial and redevelopment problems of such dimensions that the traffic itself needed to be restrained. With the growing expertise in the measurement of such externalities, this approach has been extended to the estimation of the extent of over- and under-load of traffic within a road network in respect of environmental capacity loads. Such estimates have been used as an aid in the design of traffic improvement and management measures after points of serious overload have been identified.[16]

This need to integrate the traffic and the environmental and financial aspects of land-use planning was paralleled in another direction. As the road programme led to the building of urban motorways to cater for the growing traffic it was recognised that if the location and design of such routes was determined only by engineering considerations (cost and traffic), it offended the quality of life in urban living. This led to the need to introduce route location local environmental considerations as well as those related to engineering.[17]

From quite another direction came the recognition that the concentration of transportation planning on the motor vehicle had led to serious neglect. The tendency to regard "traffic" as that in motor cars, ignoring other modes (walking, cycling, buses, commuter trains, etc) led to the neglect of those groups of the population who either preferred these other modes or of necessity had to use them (the poor who could not afford the motor car; the family with one car who could not afford more; the elderly and the disabled). This led to the neglect of facilities for these other modes (cycle tracks, public transport, etc) which hastened a decline in these modes when the reverse was needed to solve the problems caused by the motor car.[18] By the same token, the concentration on the motor car undermined the possibilities of integrated and coordinated transport within the transportation sub-system. Indeed, it was only recently that a "transport policy" was introduced to relate to all modes and not separately to roads, ports, or airports.[19]

These and similar incremental changes in the attitude and practice of transportation and traffic planning and engineering have helped to integrate the practice closer with land-use planning, which takes the broader view. Over recent years the very decline in Western economies has influenced both transportation and land-use planning, and by doing so has forced them further towards integration. The reduction in the rate of economic growth and availability of capital resources required for transportation infrastructure has meant a much more careful scrutiny of the "value for money" of the infrastructure; and the need to find low-cost solutions to traffic problems has led to an increasing regard for traffic management, that is making more effective use of a given road network with the aid of limited and low-cost improvements, such as traffic lights, with due regard to environmental constraints.[20] In parallel has grown the use of mechanisms to contain and reduce the demand for trips by motor vehicles.[21] All this has in effect sounded the death-knell of the kind of traditional transportation study which was predicated on the need to spend a vast amount of resources in roadworks; and it has meant a new approach in establishing priorities in the use of very scarce resources.[22]

HOW CAN LAND-USE PLANNING HELP
IN FACILITATING TRAFFIC PLANNING OBJECTIVES?

Since the land-use potential and transport capability are so closely interrelated, it follows that planned changes in land use will have the repercussions on traffic needs and, therefore, on the design of the transportation system. How can this help in facilitating traffic planning objectives, that is, facilitate the movement of goods and people, having regard to their trip purpose, economy in time and financial cost, avoidance of environmental pollution and so on? This question has arisen forcefully in view of the need to conserve energy following the rise in oil prices. How can changes in urban physical patterns affect transportation costs? Clearly, there will be differences in towns of high or low density; where a high degree of accessibility to local facilities avoids long journeys to a centre; where the form is linear as opposed to concentric, with consequential repercussions on the length of journeys and trips.[23]

These possibilities are very evident when considering the end state or urban form, for example, when a new town is planned. But this is the rare situation. More frequently we are concerned with the expansion of an existing town, planning to influence trends rather than build a new settlement. Here the possibilities are by definition more limited. For one thing, there is the difficulty of introducing the kind of surgical operations needed for freeing traffic in the centres of cities, so that the established urban form becomes a constraint on possibilities; for another, the extent of areal change in an established town in a developed country might be as low as one or two per cent a year, which hardly offers room for manoeuvre in affecting all the traffic activities of the town.[24]

In this situation it is not only the areal spread or renewal that is critical but also the urban trends which are proceeding. For example, in any particular town there could simultaneously be a tendency towards dispersal from the centre of those uses which find it more economic to locate on the outskirts (e.g. large factories and warehousing interchanges and a concentration at the centre of tertiary activities which value highly the traditional locations. Here a land-use policy could be followed which encouraged those trends which would minimise the traffic problems, for example by making it easier for dispersing activities to find suitable locations where their traffic generation could be catered for, i.e. where there is adequate space in a large town for parking and access. Equally, a trend of dispersal from the main centre to subsidiary centres could be encouraged, for this would permit the substitution of local trips for those needed previously to the centre.

Principles such as these were applied in considering alternative possible approaches to the traffic problems of one city in Britain, where a historic core with a medieval wall made difficult the conventional road proposals.[25] Three possible elements were considered:

□ influencing the location of activities by altering the existing pattern of land use or altering the location of new uses, with the aim of creating a corridor of movement for

certain journeys, particularly work journeys, where public transport or an adequate road can deal with the movement with minimum disadvantage to the remainder of the community, or drawing some of the demand for movement away from the existing fabric;
□ influencing the timing of activities by persuading people with different trip purposes to travel at different times; or influencing the time at which people start or finish work and shopping;
□ influencing the mode of travel, for example towards greater use of public transport, walking or cycling; as for public transport, the aim would be to make it sufficiently attractive for current users of buses to encourage them to continue using them, for motorists to encourage them to use buses rather than cars, and, in particular, for potential motorists for the same reason.

This last approach - influencing the mode of travel - was applied in a particular British new town, Stevenage.[26] Here the road planning for the ultimate expansion of the town visualised sufficient motorway capacity to provide for "full motorisation" with only vestigial transport by buses. But when the decision had to be taken whether or not to add significantly to the road system, the possibility was explored of improving the bus system in order to provide an alternative to the daily commuting to work, which was the critical peak flow. Following exploration of a design for a suitable scheme, involving buses running to fixed schedules with seats reserved for commuters and showing that there would be an overall net advantage to the Stevenage community by so doing, an experimental service was initiated. This ran for about two years and the experiment justified to a large degree the predictions made in the social cost benefit analysis.

This attempt to influence the mode of transport is only one of the means which are necessary to supplement measures for minimising traffic problems through land-use change, which are in themselves indirect (since the traffic is the objective, not the land use) and also slow to take effect. Such measures relate to pricing of the vehicles in use, for example through congestion taxes or fare differentials; by providing subsidies for public transport so that fares can be lowered to attract custom; by licensing and perhaps subsidising intermediate forms of transport, such as jitneys, mini-buses with ensured seating, shared taxis etc.[27] And then there are the more physically oriented means of offering priorities to vehicles which make joint claim on restricted road surfaces, as the bus lanes introduced in London; limiting the parking available at the destination of downtown traffic, both in the public car parks and in those for private use in particular buildings.[28]

THE NEED TO INTEGRATE TRANSPORTATION AND LAND-USE IMPACT ANALYSIS
Impact analysis is the prediction of the repercussions in a system resulting from some proposed intervention, in the form of a change in policy, of regulation, of capital investment in works, etc. As such, it has a long tradition in both the transportation and land-use fields. But these are of a different nature.

In transportation the intervention can be in one of the various forms mentioned above, but classically it is in new infrastructure works, be it in road or rail, etc. Here the purpose is to predict the changes in transportation activity which will be caused by the intervention in terms of differences in the activities of the traffic, be it in quantity, mode or length of time of trip. These are clearly the "user benefits". But under the influences which led to the broadening of horizons of transportation planning mentioned above, there has been a growing regard also for the "non-user benefits", those outside the transportation system. Starting perhaps with impacts on adjoining land (environmental nuisance, or prospects for development) concern has widened and today could also cover financial, economic, social, environmental and aesthetic impacts of the change in traffic activities flowing from the intervention.[29]

By contrast, land-use impact analysis has been broad from the beginning in order to predict the repercussions on the way of life of a community of the mixed development implicit in the preparation of an urban and regional plan. But while it is part and parcel of conventional planning practice, it has not been known under the name of "impact analysis". Furthermore, in conventional planning circles, it has not been advanced to the same degree as impact analysis practised under other heads (economic, environmental, transportation, etc).[30] The current situation is therefore this. "Impact analysis" in land-use planning has developed over a broad front and has been enriched by sectional impact analysis in a variety of skills identified with transportation, economic, social, environmental and other impact analysis. This in itself enables the principles and practice of "urban impact" analysis to be advanced.[31] But it also raises the necessity for any sectional impact analysis to be broader than the sector from which it originates. For any particular sector this will come about by tracing through the social, economic, environmental and other repercussions of the sectoral input (the road or factory) throughout all the urban and regional sub-systems.

In other words, it is necessary for the comprehensive land-use view to be taken of any sectoral impact analysis. By doing so for transportation, the integration which is the concern of this conference would be better aided.

THE NEED TO INTEGRATE TRANSPORTATION AND LAND-USE PLANS
- OR PROJECT EVALUATION

Plan or project evaluation is the comparison of the impacts from alternative interventions on the system with a view to making a judgment or decision on which of the alternatives would be preferred by a given criterion, such as the general interest of the community. The evaluation therefore is the assessment and comparison of the outputs of alternative means of intervention in terms of the differences of impacts on the on-going system. As such, impact analysis and evaluation can be seen as part of one process, "because, after all, what else is it that consistutes transportation-system evaluation, but the consideration of impacts of various transportation-system alternatives on users and non-users of the respective alternatives".[32]

But they are often seen as separate processes, with impact analysis falling short of the evaluation, and evaluation being applied to inputs which are the output of the impact analysis process. This dichotomy, for example, is clearly seen in the environmental impact field.[33]

Here again it is of interest to trace the parallels between transportation and land-use planning. In the former, the art and science of project or system evaluation is well established.[34]

But there has been the tendency noted above to confine the evaluation to the impacts for road users, with only slow and partial extension to the non-users. By contrast, the assessment and evaluation of alternative plans has long been implicit in land-use planning, where it has by definition embraced all sub-systems in the urban and regional system and all users and non-users - at least in principle. The methodology has been limited in practice, however, and it has emerged only comparatively recently.[35] And, as with impact analysis, the methodology has been enriched by contributions from other disciplines.

Here again, therefore, the evaluation of transportation follows the path of impact analysis in veering towards an assessment covering much wider sets of impacts than the traditional transportation benefit-cost analysis. Extension of this kind from the traditional has recently been advocated in a report relating to trunk roads in Britain.[36] The methodology advocated takes in not only the differences in costs and benefits attached to the various impacts, but also the incidence of those costs and benefits on various groups which would be immediately affected.

Evaluation of such wider scope is assisted if in preparing the transportation proposals there is an awareness of the considerations which should be taken into account in the evaluation: that is, where evaluation criteria become design criteria.[37] Such an approach is seen in one recent planning study which recognised that the studies should be built around issues on which there has been public concern, namely objectives which the transport plan would seek to fulfil, standards which reflected conditions that would be generally acceptable to people, and problems which were revealed both by comparing the existing situation with the defined standards and by examining what the various groups within the community regarded as problems.[38]

This need to integrate transportation and land-use planning in the realm of evaluation is of more than academic interest, simply because it is on such evaluation that the decisions on alternatives are taken. Some instances can be given where transportation evaluation which was too narrow in its framework would lead to clearly erroneous conclusions.

In developed countries which are blessed with wide-ranging railway services, the product of the railway age where neither the motor car nor the aeroplane offered an alternative, some railway lines in the rural areas are running at a loss, simply because of a fall in demand and a rise in operating costs. But the question arises: should the operator cut out the line in question on financial grounds? An investigation based on a narrow foundation (costs and revenues, or cost and user benefits) could well lead to elimination. But when the passengers would not have a ready substitute, such as a motor vehicle, private or public, the losses would clearly be underestimated in the option for elimination. This could be particularly so where the traffic in question is not merely residentially based but relates to, for example, tourism or holidays in the peak season.

Similar reasoning can apply to public transportation in cities (e.g. buses or trams) where the criterion for improving or diminishing the service rests on costs and returns on cost to the operator, and benefits to the user. Here the position is worse than in the rural areas, if only because the presumed alternative, the motor car, is often not realistic in that there are inadequate roads for the compensating increase of motor vehicles. Thus there could be losses from closure on two sides: bus users who shift to motor cars, and bus users who cannot. There could also be further repercussions in terms of difficulties of commuting, with consequential repercussions for the supply of labour to the undertakings in the centres of cities.

Whereas the previous two examples relate to the cutting of services, the third relates to a programme of proposals. If in a city the programme is judged on a narrow foundation, a wide array of benefits and disadvantages could be ignored which would tip the balance in assessing priorities. This arises because those using the transport service do so for the benefit of the end of the trip, not for the travel itself. Thus there need clearly be no correlation between priorities based on the narrow foundation and those arrived at when the wider repercussions are taken into account. A particular "uneconomic" proposition could be very important in opening up an area for social and economically useful development, as in deprived inner urban areas. In all this arises one particularly thorny problem - that of evaluating the transport effect of alternative plans which contain variations in both land-use (by quantity and location) and transport networks. This is a familiar situation but one for which conventional transport appraisal is ill-equipped since it is based on an assured "fixed-trim matrix" (i.e. the same zone to zone trip distributions for all plans with the same allowance for generated trips). In cases where the total volume of trip making and its distribution vary between plans, it is necessary to consider prospective differences between plans in level of trip and benefits obtained, as well as in other costs. Complex formulae to represent the comparative changes in consumer surplus have been put forward,[39] making use of data derived from a conventional gravity traffic model. As an alternative, which to a large extent avoids criticism or scepticism about the predictive ability of gravity models (and anyway does not require a facility to handle highly elaborate formulae), a useful approach is to make comparisons in terms of the number or level of opportunities (say jobs, retail facilities, or labour supplies) available to households and firms at various feasible trip costs. These may then be used to supplement estimates of comparative aggregate user travel costs to permit an overall judgment on the net effect on users. These effects can then be weighed against plan differences in facility construction and operating costs and impacts on the physical environment.[40]

SOME CONCLUSIONS FOR DEVELOPING COUNTRIES

In the discussion so far there has been little differentiation between the developed and developing countries, since it has not been practicable to particularise on place or time. But there are differences in treatment between countries and cities at differing stages of development. We conclude by noting some important ones.

One essential difference is brought out above: whereas in the developed world, with a slow population and economic growth rate, the marginal change in urban expansion is slight (perhaps one-half of one per cent), the corresponding change in a city in the developing world can be dramatic (perhaps 10 per cent), e.g. São Paulo, Mexico City, Seoul.[41] This offers the advantage that the manipulation of changes in land use can produce more significant results for the current urban area than for those in the developed world. For example, there is the opportunity to influence urban form more dramatically than with the slow rate. There could arise plans for a poly-nucleated town (as opposed to one which is heavily centralised), with the advantage of facilitating shorter trips to local centres, and avoiding a concentration in the main centre which is often too congested to absorb it. When this consideration is related to the general lack of resources and large sectors of low income, there is clearly some need to consider all forms of transportation, not only that of the private motor car which can be afforded by only a minority, at least for a considerable period ahead. Thus, if attention were to be given to the alternative and "suppressed" modes, not only would there be greater benefits in transportation services for a wider section of the population, but habits and facilities could be initiated which would head off too dramatic a push towards modernisation.

In some developing countries (Korea is an example), however, rapid growth in population and per capita incomes in the past five to ten years has caused very rapid growth in the number of vehicles in the cities, resulting in congestion in city centres and preventing what public transport services there are from operating efficiently. In cases such as this, the opportunities for taking explicit account of the relationship between land-use and transport planning are perhaps the greatest, given the combination of available resources and the will to tackle the problem in a carefully planned manner. In an attempt to find solutions to the Korean city problem, a recent city study investigated the following strategic policy alternatives: manipulation of demand for private cars by taxation of ownership (vehicle purchase taxes) and taxes on car use (e.g. petrol taxes); the provision of alternative modes of transport, which should be encouraged, ranging from a public bus service to the city car's closest rival, the private taxi; and the establishment of urban land-use forms consistent with the first two alternatives. The alternatives investigated included the development of satellite towns to absorb expanding populations and the expansion of the existing city's limits.[42]

This then is a clear example of integrating transportation and land-use planning and policy in what must represent the major challenge in metropolitan planning today: the burgeoning cities of the Third World. And it is here that the greater pay-off of such integration can be seen.

REFERENCES

[1] For a historical account see Cherry, G., Urban change and planning: a history of urban development in Britain since 1750 (Foulis, 1972).

[2] United Nations, Report of Habitat: United Nations Conference on Human Settlements (UN, 1976), Chapters 1-3.

[3] Lichfield, N., and Marinov, U., "Land use planning and environmental protection: convergence or divergence", Environment and Planning, Vol 9, No 8 (1977), pages 985-1002.

[4] For a recent prediction see Martin, J., The wired society (Prentice Hall, 1978).

[5] For example, Blundell, W.P., The land-use/transport system: analysis and thesis, (Pergamon Press, 1971), Chapter 1.

[6] Adapted from Lichfield, N., Kettle, P., Whitbread, M., Evaluation in the planning process (Pergamon Press, 1975), Chapter 2.

[7] See, for example, Nickson, J.V., and Batey, P.W.J., "The analysis of transport within a metropolitan county structure plan: the Greater Manchester Experience", Town Planning Review (July 1978).

[8] Shunk, G.A., and Turner, C.G., "Changing goals of urban transport planning", Proceedings of Annual Transportation Board Conference (1975); and Turner, C., "A model framework for transportation and community plan analysis", Journal of American Institute of Planners, Vol 38, No 5 (1972), pages 325-31.

[9] See Webber, M.M., "The BART experience: What have we learned", The Public Interest, (Fall 1976).

[10] Lichfield International Incorporated (Beesley, Edwards, Turner, Gist, Kettle), BART impact programme, Land-use model project, Comparative land-use impacts of rapid transport (US Department of Transportation, 1978).

[11] Lichfield, Nathaniel, and Partners, Seoul metropolitan transport study: Technical memorandum No 5: Measurement of social costs and benefits and associated data requirements (1978).

[12] Schaffer, F., The new towns story, (London, MacGibbon and Key, 1970).

[13] Ministry of Transport, Roads in built-up areas (HMSO, 1946).

[14] Tripp, H.A., Town planning and road traffic, (Arnold, 1942)

[15] Traffic in towns: report of the steering group (Buchanan Report) (HMSO, 1963).

[16] Buchanan, Colin, and Partners, Alternatives for Edinburgh: second interim report of Edinburgh planning and transport study (1971).

[17] Report of the Urban Motorways Committee (HMSO, 1972).

[18] Hillman, M., Henderson, I., and Whalley, A., Transport realities and planning policy (Political and Economic Planning, 1976).

[19] Department of Transport, Transport policy (Cmnd 6386, HMSO, 1977).

[20] Ball, R.R., "The TPP: assessing priorities", The Municipal Engineer (1977).

[21] Plowden, S., Towns against traffic (André Deutsch, 1972), Chapter 9.

[22] Thomson, J.M., Method of traffic limitation in urban areas, (OECD, Division of Urban Affairs, 1972).

[23] Lichfield, N., and Lichfield, D., "What change is needed in planning methodology?", Proceedings of First International Conference on Energy in Community Development (1978, to be published).

[24] World Bank, Urban transport: sector policy paper (1975), Chapters 1-2.

[25] Lichfield, N., and Proudlove, A., Conservation and traffic (Sessions Book Trust, 1976) Chapter 13.

[26] Lichfield, Nathaniel, and Associates, Stevenage public transport: cost benefit analysis (Stevenage Development Corporation, 1969).

[27] World Bank, Urban transport: sector policy paper (1975), Chapter 34.

[28] Thomson, J.M., Method of traffic limitation in urban areas (OECD, Division of Urban Affairs, 1972)

[29] Stopher, P.R., and Meyburgh, Transportation systems evaluation (Lexington Books, D.C. Hearth and Co, 1976).

[30] McEvoy, J., Dietz, T. (eds), Handbook for environmental planning: the social consequences of environmental change (John Wiley, 1977).

[31] Such an urban impact analysis is being now required in all HUD federal programmes in the USA (US Office of Management and Budget Circular A/116, 16 August 1978).

[32] Stopher, P.R., and Meyburgh, Transportation systems evaluation (Lexington Books, D.C. Hearth and Co, 1976), page 97.

[33] Lichfield, N., and Marinov, U., "Land use planning and environmental protection: convergence or divergence", Environment and Planning, Vol 9, No 8 (1977), pages 985-1002.

[34] Beesley, M.E., Urban transport studies in economic policy (Butterworth, 1973), Part 1.

[35] Lichfield, N., "Evaluation methodology in urban and regional plans: a review", Regional studies (1970).

[36] Report of the Advisory Committee on Trunk Road Assessment (Leitch Report) (HMSO, 1977), Chapter 20.

[37] Adapted from Lichfield, N., Kettle, P., Whitbread, M., Evaluation in the planning process (Pergamon Press, 1975), Chapter 1.

[38] Sharman, F.J., Road planning in a changing society (Tenth Rees Jeffreys Memorial Lecture) (1978, to be published).

[39] Neuberger, H., "User benefit in the evaluation of transport and land use plans", Journal of Transport and Economics and Policy, Vol 5 (1971), pages 52-75.

[40] West Midland Regional Study: Technical appendix 5: Evaluation 2 (July 1972).

[41] World Bank, Urban transport: sector policy paper (1975), Chapters 1-2.

[42] Beesley, M.E., Turner, C., and Gist, P., (Nathaniel Lichfield and Partners), and Whang, I.B., (Korean Institute of Science and Technology), Korean urban sector study: options for secondary city urban transport (1978, unpublished).

The 21st century metropolis: a land-use and transportation perspective

2

Herbert S. Levinson   United States

SYNOPSIS

This paper looks at the 21st century American city in the context of current trends and public policy options. It shows how various land development strategies, social programmes, transportation technologies and energy supplies will affect urban forms, life-styles and mobility. It describes the steps needed to bring about major changes in existing development patterns and trends. It presents various futures in the form of scenarios for twenty-five, fifty and seventy-five-year planning horizons.

INTRODUCTION

Cities throughout the world continue to grow in size, complexity and importance as economies strengthen and populations expand. Managing this future urban growth to maintain mobile and livable environments will remain an important challenge in the years to come.

There is no simple answer. Cities of differing size, culture and economy will have their own specific needs, both in the developed and developing countries. Bangkok and Boston, Jakarta and Jerusalem, Tehran and Toronto, each will have its unique set of land-use and transportation requirements.

Relations among nations, the stability of governments, and rates of economic growth will have an important bearing on the future metropolis. Other key factors include population expansion, willingness to renew old cities and build new ones; energy availability; new systems for acquiring, allocating, managing and servicing urban land; energy and resource availability; and continued advances in technology and governmental structure.

Predicting how these factors will interact is no easy task. It calls for understanding the past, assessing the present and probing the future to find ways to bring resources, society and technology together.

The observations which follow focus on the North American metropolis. They show how space, time and public policy will interact in shaping urban environment. In varying degrees, they may also apply to other cities in the developed world.

A HISTORICAL PERSPECTIVE

One way of viewing the future American metropolis is by "temporal analogy", i.e. looking at changes which have occurred over comparable time spans in the past, to answer the questions: How much different will the next twenty-five years be from the past? Will it change more or less rapidly over the next century than it did over the last? What qualitative differences will take place?

<u>Technology</u>. We are all aware of the effects of technology on urban form, for our cities are largely products of modern industrial society. The roles of the skyscraper, automatic elevator and electric railway in centralising cities are well documented. The motor vehicle, refrigerator, air conditioner and new industrial processes are among the technologies that have played equally significant roles in recent decades. They, too, have influenced the size and form of cities, the internal distribution of population and activities, and the modes of travel. The modern jet aircraft has vastly improved air travel and today the airport is often the second largest urban activity centre. Tables 1 and 2 show some of these interrelationships.

Between 1875 and 1900 the steel-frame skyscraper, vertical elevator and electric railway changed the scale of the city and its centre. The "electric railway city" replaced the "pedestrian city" as the radius of development increased from three to over ten miles. This city reached its peak by about 1925, prior to the great depression.

Between 1925 and 1950 urban change was in many respects less dramatic. The trend towards decentralisation continued, although it was constrained by depression and war.

Between 1950 and 1975 the "automobile city" emerged. There was decentralisation of shops, industry and housing; expansions of suburbs and decline of the central city; and a shift from public to private transport. Problems of central city obsolescence and tax base; urban sprawl and declining transit passenger numbers; environmental intrusion and energy consumption led to accelerated federal programmes with a highly varied urban response.

Between 1975 and 2000 "new technologies" will mainly improve the "efficiency" of existing technologies. (For example, automatic train operation on new rail transit systems, computerised traffic control systems and modular housing developments.) Communication is not likely to reduce the need for transportation or direct personal contact.

Future advances in urban technologies, such as development of viable low-cost personal rapid transit or people-mover systems, data automation and computer advances, new building techniques, and new goods distribution methods, will influence future urban activity patterns. Because new technologies, particularly those concerned with transportation, evolve gradually over many years, no radical changes during the next several decades are expected in current travel modes, movement patterns or urban forms. Moreover, the existing urban environment serves to constrain the rapid introduction of new transport technologies.

<u>Social and economic factors</u>. The second world war triggered a major influx into metropolitan areas of people who previously had lived in rural areas. This last "great wave of immigration" initially was caused by job opportunities in war plants and, subsequently, the attractiveness of urban welfare programmes in major urban centres. In turn, the influx of rural and minority people into the central city caused social unrest and resulted in middle-class whites leaving the central city for the suburbs.

The period after the second world war reflected the preferences of climate, space and privacy. Suburbia was largely a response to the desire for free-standing houses, informality and, in some cases, proximity to outdoor recreational resources. FHA low-cost housing loans and improved construction equipment also contributed to the rapid suburban growth.

Planning initiatives and public improvements. Major urban planning efforts, such as the L'Enfant and McMillan plans for Washington and Burnham's plan for Chicago, have left their mark on many of our major cities. So, too, have many bold public improvements which have become a reality. The impacts of New York City's subway system and its Grand Central and Penn Station terminals are well known. Equally significant are the bridges, beaches, parks, and parkways defined in the First Regional Plan which became a reality some 25 years later. Almost every metropolitan area has public improvements which have influenced urban form and amenity. Examples are Chicago's lake-front development, New Haven's new-town centre, Los Angeles' freeways, and Washington's in-town south-west renewal. Viewed in this context, the 21st century metropolis will reflect those current plans and projects which will become a reality by 2000, for example the Dallas-Fort Worth regional airport.

Past perceptions. Planners and urbanists of the past also provide insight into the future, especially the extent to which their proposals and visions have been realised. Frank Lloyd Wright's Broadacre City, a decentralised community where the distinction between country and city is obliterated, has been achieved, even though not in Wright's organic form.[1] So, too, has Le Corbusier's Radiant City, his blueprint for the future, although in a much less qualitative way.[2] In reality, the urban development process is far too complex to fit any such concept.

GROWTH TRENDS - 1975-2000

In the years to come urban settlement patterns will, of course, be influenced by relationships among countries and the national economy. Barring war, sustained oil embargoes or major economic recession, the following trends are likely within the United States:

The nation's population will continue to grow. The dimensions and patterns of this future population growth are well known and documented. There will be fewer young people (at least in the decade ahead) owing to a declining birth rate, and more older people owing to greater longevity. The nation's 1977 population has been estimated at 217 million and 1975 census projections anticipate 245 to 287 million by 2000, compared with 271 to 322 million projected in 1971.

Metropolitan areas will continue to grow and continue to absorb most of the national population growth, except perhaps for a few large centres in the north-east. Most of the growth will take place in exurban and suburban areas. "Megalopolis" will have its Great Lakes, Pacific Coast and Gulf of Mexico counterparts. There will be growth in smaller towns and cities because of a better quality of life, more recreational opportunities, and increased job opportunities resulting from communications and aviation improvements.

"Sun-belt cities" will continue to grow rapidly while there will be much slower growth in the north-east, owing to the continued regional population shifts from the north and east to south and west.

Urban living and travel preferences will reflect continued economic growth. The number of two-wage-earner families will continue to grow as more "working wives bring home a second income". In the future, as in the past, rising incomes will raise per capita space requirements, although probably at a lower rate than in the past decades. Almost every comprehensive urban transportation planning study has found that urban land consumption and travel outpaced population growth over the past several decades. Unless constrained by public policy or a sustained shortage of energy resources, these trends will probably continue.

The changing mix of the nation's population will bring new types of housing to the urban scene. "Planned unit developments" combining single and multi-family residential units will become increasingly common in response to changing community needs and values, as well as the increasing costs of residential construction. There will be new developments for specific age groups, such as retirement communities for senior citizens. An increase in leisure time, such as might result from a four-day work week, will accelerate the growth of second homes, vacation, travel and outdoor recreation. Second homes may be as far as three hours' travel time from home, provided there are no travel constraints. Apartment and condomimium living will become more popular as the proportion of elderly in the population continues to increase, and there are more later marriages, childless marriages, unmarried couples and divorces. For some, rising fuel prices and disenchantment with home ownership will also make such a life-style attractive.

Decentralisation of commercial and industrial activities will continue. Despite active downtown building programmes in many cities, and despite concern over air quality and energy, the number of regional shopping centres continues to grow. New industrial developments are mainly keyed to highways and airports outside the built-up cities, and the corporate office park has now emerged as a challenge to the city centre.

The city centre will grow slowly in most urban areas. Growth will take place mainly in government, public facilities, services and finance, but in some large cities there will also be growth in the retail, hotel and residential sectors.

Urban person-travel will continue to outpace population growth. The greatest increases will be in non-work trips. Work trips will probably grow from 0.6 to 0.8 trips per capita per day. By 2000 there will be about 16 daily person-miles per resident, compared with about 12 today. Trips to the city centre will be longer than today. Most travel growth will take place in suburban rings around the city, often as "circumferential" trips. Automobile travel will continue to grow and there will be some gain in public transport trips.

Public policy will influence urban settlement patterns to the extent that is real, responsive and timely. Evolving federal policies and programmes relating to national urban growth, urban housing, renewal and land-use, transportation funding, energy, and the environment will progressively influence urban growth. Recent years have seen a decline in urban freeway construction and rethinking of public transport strategies in view of limited resources and changing priorities. Urban renewal efforts appear to be accelerated or retarded in accordance with specific federal attitudes. Federal (or state) financial aid to cities and participation in urban welfare programmes will affect impending fiscal crises.

A new federal energy policy will influence urban form and transport, especially in the long term. The shift away from natural gas is already apparent. In the short term, motor vehicle travel may be suppressed, particularly long-distance recreational trips. Eventually an energy-conservant city may emerge, as urban development decisions adapt to energy availability and as compactness replaces spread.

These trends imply several important planning challenges for the urban area and its central city. Improved government arrangements, economic incentives and land-use controls will be needed.

Planning for growth implies careful emphasis on regional development patterns, and on establishing those key planning decisions which will affect regional growth, i.e. water supply, utilities, access, open space and employment.

Planning for little growth (or "zero population growth", although this is perhaps not realistic) calls for qualitative improvements within the existing urban area, i.e. better housing, community facilities and transportation, and improved social, economic and educational programmes.

We have taken many important steps to preserve or revitalise our city centres. Continued action by government agencies and private developers will be necessary, and can be expected. The improvement opportunities are highly selective, however, especially for retail and residential activities. In-town residential development offers most promise, where amenities can be improved (i.e., Inner Harbor, Baltimore) or where the area is not surrounded by declining residential neighbourhoods. The extent of downtown rebuilding will depend on the specific economy and situation of the city centre, and initiatives of the public and private sectors. There is no magic formula.

There is an urgent need to upgrade the areas surrounding the city centre. This is a far more challenging task, since there are no obvious solutions. Almost every city has such declining wholesaling, warehousing, manufacturing and railroad areas. New modern industrial parks, in-town college or vocational school campuses and, in some cases, "new towns in town" are among the opportunities which should be explored. Finding and implementing the right mix of activities and economic incentives is an important planning challenge.

Within the central city, there is an obvious need progessively to improve housing stock, schools and community services. Over the next several decades, similar renewal efforts will also be required in many of the older suburban areas which surround the central city.

THE 21st CENTURY METROPOLIS
The preceding trends and needs provide the context for various scenarios for the 21st century North American metropolis, for how well we plan for change today will profoundly affect the quality of urban life tomorrow.

One approach is to define metropolitan options in terms of alternative land development strategies for a given time. Such land-use options could include the traditional "trend or sprawl", "multi-centre" and "corridor" development concepts which vary key decisions such as downtown employment, central city renewal, public open space, population distribution, highways and public transport, and public utilities.

A more realistic approach would be to vary the extent of public intervention and technological change through time. This approach leads to the following three views of the 21st century metropolis.

Metropolis 2000. This will be one of "minimum change". It will be generally similar to today's metropolitan area, if somewhat larger and more dispersed. The reasons are apparent. Much of this metropolis is "in place" today, and little change is expected in built-up areas. Moreover, there is a time lag between the initiation of public policy and its translation into actual changes in the urban landscape. In many respects, we are looking at this metropolis today.

The city centre will grow as a result of continued renewal efforts, but mainly as an office and government centre. There will be some upgrading of the urban housing stock, but the "rings" around the city centre will continue to decline, with a net loss in central city population. Older suburbs, too, will have problems of form, obsolence and, in some cases, economic survival. Suburban shopping malls, industrial parks and office clusters will continue to proliferate, but at a declining rate.

There will be improvements in the quality of public transport services, and some gains in passenger numbers. Rapid transit, or "LRT" will have become a reality in cities such as Atlanta, Baltimore, Buffalo, Detroit, Los Angeles, Miami and Pittsburgh, and we will be able to assess its impact on urban structure and land use. Selective freeway construction and better traffic management will improve road travel.

Rising petrol prices and potential problems in fuel availability will limit future increases in automobile travel, although the car will remain the primary travel mode for most urban trips. They will first restrain long-distance vacation and weekend recreational trips and then the journey to work as well. Trip sharing (multiple stops per car) will be more common than ride sharing (more persons per car).

We will have learned many lessons by the turn of the century, which will lead to better cities in the years ahead. These are that the proliferation of city and suburban governments may not be in our best interest; that cities are too important to abandon; that increased commitment to renew our cities in a physical and social sense is vital; that land as well as energy is a scarce resource; and that urban transportation system development calls for a creative partnership with urban land development.

We will have learned how to recognise fads and place them in clearer perspective. Examples are the following: first, the focus on long-range planning, and then on short-range planning; that freeways are vital, then that they are in conflict with cities; that rapid transit is essential, then that it is unnecessary; that mobility can be improved only by construction, then only through operational improvements (traffic management). We will have learned how to put the pieces together in a more rational manner.

Metropolis 2025. This will reflect the result of more than a quarter of a century of concerted public action. The models of metropolitan government, found today in Toronto, Miami and Indianapolis, will have many counterparts in other metropolitan areas. We will have found means to encourage the flow of private investment into our metropolitan areas, to finance our urban communities through private capital and through municipal, state and federal funds.

We will have implemented new metropolitan approaches to housing, education and economic development, in addition to land-use and transportation. The city centre will once again have 24-hour life, and the ring around it will contain productive commercial, industrial or residential areas. We will have upgraded our urban housing stock. More importantly, we will have initiated the essential social and economic programmes which will bring improved opportunity to all urban residents.

The revitalisation of our cities; the clustering of people and jobs along transportation corridors; selective increases in urban densities; and the expansion of public open space will improve urban form, increase urban compactness, reduce travel distances, make public transport viable, and conserve energy.

There will be more "off street" rail and bus transit which will be built at minimum capital and social costs, commensurate with demonstrated need. Strategically developed motorways will complement public transport in providing mobility and shaping urban growth. Within the centres of our large cities, selective automobile restraint will become a reality not only through improved public transport and energy constraints, but by the redesign of the centres themselves to a pedestrian scale.

Metropolis 2050. This will reflect the impacts of new technology as well as public policy. National urban growth, brought about by the needs to conserve energy and equalise opportunity, will limit growth in large metropolitan areas and create new communities.

The new cities of the 21st century will be built in an environmentally sensitive and energy-conservant manner. They will be compact, yet diverse, and afford a wide range of housing and employment choices. They will cluster development and transport in order to minimise the need for travel. Public transport services, private transport vehicles and pedestrians will each have their own separate rights of way in a full application of the principle of access control.

These new cities will be built around new construction, energy and transport technologies. Personal rapid transit may become a dominant travel mode, and palletised conveyor systems may move urban goods. Nuclear and solar energy may reduce dependence on petroleum-based energy sources.

Existing cities, too, will enjoy the benefits of new technologies, yet they will retain their charm and heritage of the past. They, too, will embody our visions and plans.

THE METROPOLITAN CHALLENGE

These views of tomorrow's metropolis identify some of our future choices in the United States. The challenge is for increased commitment in the years to come to anticipatory rather than reactive planning. It is time to plan now for tomorrow's metropolis, to take the necessary bold actions to improve the quality of urban life. In this way, our cities will continue to be the centres of our culture and society.

REFERENCES
[1] Wright, F.L., The living city (1963), Mentor Books.
[2] Le Corbusier, The radiant city, first published in 1933 under the title La Ville Radieuse; published by Orion Press, New York, 1961.

Table 1. An overview of urban growth, technology and public policy

Year	Type of city	Technology	Prevailing urban transport	Public policy factors	Economic and political factors
1875	Pedestrian	Steel frame construction Electricity (DC & AC motors) Vertical elevator	Cable-car		
1900	Electric railway	Subway Gasoline engine (Automobile)	Electric street car	Federal aid	World War I
1925		Radio Air conditioning Diesel engine Freeway		FHA-housing	Immigration limited Great Depression World War II
1950	Automobile (Urban region)	Nuclear development Television Computers Jet aircraft Space travel	Motor bus & automobile	Interstate Highway system Urban Mass Transit Act	
1975					Energy crisis

Table 2. Transport mode and urban form

Item	Type of city		
	Pedestrian	Electric railway (rapid transit)	Automobile
Population	3,000,000	3,000,000	3,000,000
Area (m^2)	30	200	400
Density (persons m^2)	100,000	15,000	7,500
Jobs in city centre	200,000	300,000	150,000
Development pattern	Compact	Major corridors	Dispersed
Example	Paris, 1900	Chicago, 1920	Los Angeles 1970

The four transitions from traffic engineering to environmental planning

3

Hans B. Barbe   Switzerland

SYNOPSIS

The traffic engineering profession is quite young, compared with other sciences such as medicine, law or philosophy. Yet, in its short lifetime of hardly half a century, it has undergone a fast and creative development wich is characterised by a number of transition phases. In the first transition, the plain engineering aspect was replaced by a broader planning outlook. It was recognised that it would hardly be sufficient to administer remedies in order to arrive at a satisfactory traffic structure and that some creative thinking and forward planning were needed to adapt traffic fully to the requirements of modern urban life. But soon the next transition was under way. It became evident that the former restriction of the profession to motor vehicle traffic was not adequate and that transportation should be understood as an overall process. This development eventually led to renaming the ITE the Institute of Transportation Engineers in 1976. After a while, the next transition emerged from the fact that transportation is only one sector in the wide spectrum of human activities in our society. To understand transportation needs properly, other aspects have to be taken into consideration as well: economics, recreation, job and dwelling distributions, operation and management. Thus, integral planning developed as an interdisciplinary methodology. Finally, in the early 1970s the increasing worldwide concern for environmental problems triggered the latest transition, from the integral planner to the environmental planner who, of course, is still as integral as before, if not more so. His responsibility now comprises not only his particular clients, his society or his region, but also our environment, other regions and later generations. The conclusions to be drawn from this development are that the transportation planner must remain versatile and flexible. He should have a strong sense of "citizenship" to fulfil a responsible role in a fast changing society. But the same holds true for those in charge of awarding contracts and implementing them. There is no place for narrow-mindedness in our time.

INTRODUCTION

It may be with a certain envy that the traffic engineer regards his colleagues in other fields of learning. Doctors, lawyers or philosophers can look back on a history of centuries, if not millennia, of homogeneous and fairly continuous development. Hippocrates, in whose name young doctors still pledge themselves to ethical conduct, practised 2400 years ago. The principle "in dubio pro reo", still one of the fundamental requirements of civilised justice, was already a guideline of Roman law 2000 years ago.

Compared with these time-scales, our traffic engineering profession seems young and undeveloped indeed. It may be some small comfort to know that the present standard railway gauge is in fact based on the distance between the wheels of Roman chariots, which were standardised because otherwise they would not have fitted into the grooves cut into the paving blocks. We also know that one-way schemes and parking restrictions were already in force in the Roman Empire, but all this does not mean that traffic engineering is an old and established science.

Yet, it is surprising how many transitions this profession has undergone in the short half century of its existence. These transitions have obviously not yet reached their final stage and are still in the process of maturing and establishing a young and growing discipline. They will also help to ensure that the contribution of the traffic engineer to his society will meet the urgent requirements of the time, requirements which change as rapidly as traffic and transportation themselves. So far, it can be said that at least four major transitions have taken place.

TRAFFIC ENGINEER TO TRAFFIC PLANNER
When Harry Neal in Ohio was appointed the world's first traffic engineer in 1921, traffic in our sense, that is motor traffic, was still in its infancy. The main task of the traffic engineer was to help maintain a controlled and safe operation, mainly by introducing adequate traffic management measures. Traffic lights were installed, one-way schemes were created and uniform traffic signals agreed. Many of these traffic engineers of the first generation were electrical engineers. Their original contracts with traffic having been in the course of designing traffic signals, they had become aware of the complex dynamic problems of traffic flows, the interrelation between capacity and demand, the statistical behaviour of large volumes of vehicles (and of their users) and eventually some of the laws which govern, in particular, the urban traffic pattern.

In the United States this profession soon became a recognised branch of the engineering discipline, and in 1930 the Institute of Traffic Engineers was founded. In other countries, traffic engineering was not considered a serious science until two or three decades later. In Switzerland, for instance, the first lectures in traffic engineering started as late as 1951, and even today traffic engineers are civil engineers who have specialised, for their last year only, in this field rather than in structural design, road construction, hydraulics or others.

Eventually, it became evident that the improvement of intersections, the extension of traffic handling capacities and the continuous battle against accidents, important as they may be, do not constitute the entire problem. It began to be recognised that traffic may exert, through its own laws, an influence on the urban structure, and that accessibility had become a major factor in the selection of appropriate sites for dwellings, industries or recreational facilities. Initially, this development just happened - it was called "organic growth". After a while, however, it was found to get out of hand and planning controls were adopted to curb undesirable hypertrophical developments. The land-use planner came to the fore, generally with a basically architectural background. There was little contact or mutual understanding between his discipline and that of the traffic engineer.

In the late 1950s it was realised that there was a strong interdependence between the roles of traffic engineers and land-use planners and that they should work together.

The land-use planner had to accept that traffic was the most powerful tool at his disposal, and that the best of his plans were quite unfeasible if they were not adequately supported in terms of accessibility; whereas traffic engineers found out that they had a grave responsibility towards society through the influence of their schemes on the further growth of the towns. The problem grew beyond the question of how much capacity should be provided in a given location. The crucial question sometimes was whether or not to provide any capacity at all. Thus, the traffic engineer gradually moved into the field of overall planning and he became a traffic planner.

The comprehensive traffic plans originating in the late 1950s were the first results of close cooperation between land-use and traffic planning. Planning means that objectives of a conceptual kind are formulated - visions, as it were, of desirable long-range developments; and the means and measures which it is suggested will make it possible to implement these ideas. Soon, it became apparent, however, that something was missing, and the next transition was under way.

TRAFFIC PLANNING TO TRANSPORTATION PLANNING

In Europe, sooner than in the United States, it became evident that traffic did not consist of motor vehicles only. Public transport was rediscovered. In fact, in many European cities it was never as much neglected as in the US, because the degree of motorisation was still much lower, and the narrow geometry of historical town centres did not lend itself to the space-demanding motor vehicle. European traffic engineers have thus always had to face the problem of public transport, and they have always been aware of the importance of modal split because conditions have placed special emphasis on this perspective. Initially, because there were not enough motor vehicles to accommodate the entire volume of travel demand, and later, when motor vehicles had become more numerous, the severe lack of space helped to keep things within reasonable proportions. So, it was by good fortune rather than by special insight that the integral transportation planner developed quite early in Europe.

In the early 1960s, therefore, integrated transportation plans began to evolve. They were based on the assumption that a certain number of movements - person trips - had to be provided for between points A and B, and that this could be done in different ways. Proposed strategies differed not only in technology but also in their economic and other consequences, and balanced plans were developed which would optimise the benefits derived from a given level of investment. The fact that funds for such plans and their implementation were always quite scarce on the Continent may have helped in devising pragmatic and comprehensive transportation schemes.

The coming of the computer helped to improve the assessment of alternative strategies. It was possible to indicate the probable consequences of alternative policies and to study their feedback effect on land-use developments. It became almost fashionable to travel by train or by tramway, and public opinion soon came to agree with this philosophy. It is surprising to learn, for instance, that in Zurich only 20 per cent of all visitors to the CBD go there by car, while the remainder patronise public transport. Some of the world-famous gnomes of this city are reported to have been seen in the street-car, and if you want to meet the President of Switzerland informally, it may suffice to board the proper bus in the capital city of Berne. However, the present wave of terrorism may change these modal split conditions for obvious reasons.

Recently, this development was recognised also by the ITE. Its name was changed to the Institute of Transportation Engineers to demonstrate that traffic is only one sector of 20th century mobility. Even transport is but one sector, however, and so the next transition evolved.

TRANSPORTATION PLANNING TO INTEGRAL PLANNING

The importance of transportation is a consequence of the structure of our society. In medieval times, the artisan had his shop on the ground floor, his workshop in the basement and his living quarters on the first floor. His apprentices would occupy the attic, and commuting was a vertical rather than a horizontal movement. The cities were of pedestrian size; errands could be carried out from any place in town to any other, simply by walking there within a few minutes.

Our exploding cities are struggling with two particular modifications of this picture. First, industrialisation has led to oversized conurbations where distances are such that they require mechanical means of transport. Second, the division of labour itself creates the need for more transport movements. Responsible transportation planning could not restrict itself to registering the apparent needs; it had to start asking for the reasons why. Distorted conditions could easily result from obsolete taxing laws, local subsidies or just historical habits. From a national cost-benefit point of view, then, the best solution was not always to improve the transportation system. Sometimes, considerable investments could be saved by introducing new policies to develop certain regions or employment sectors. Integral planning had become fashionable.

In the integral planning procedure, traffic and transportation are but one segment, although a very important one, not because transportation engineers are traditionally tending to overestimate their own importance, but because transport is still a very decisive tool in all planning implementations. Without adequate transportation improvements the best ideas will have difficulty in getting off the ground; just as the best development programme can be paralysed by implementing a wrong traffic scheme.

Integral planning has grown beyond the scope of a single specialist - it is of its very nature an interdisciplinary activity. From economic to financial considerations, from recreation to tourism, from urban planning to highway design, from railway operation to traffic regulation, a number of widely differing specialities must appear on this stage of modern planning. Because many specialists find it difficult to think beyond their particular line of specialisation, the generalist or all-round planner comes into the picture to provide overall management, coordination and mutual dialogue. The questions are more complex than ever before, and systems analysis has to be introduced to bring together the divergent aspects of the many contributing disciplines.

To be truly integral, this planning procedure must also incorporate public participation of some sort. In some countries, this is institutionalised by law, in that voters are able to voice their opinions regularly at the polls. In others, public hearings are commonly used to find out whether a given plan has the approval of the citizens. It is not always certain that full democracy produces the best answer to technical questions, but we are, after all, working with the taxpayer's money and this entitles him to voice his opinion. If he is obviously wrong, this is frequently the result of insufficient information. It is most probable, therefore, that public participation will have to take the form of a system whereby a large body of representatives can be brought up to the same level of information. Experience shows that such a body can develop a much better understanding of the problems involved and that it will decide along rational and sensible lines, rather than by prejudice and emotional catch-phrases. But then a new problem arises: how these representatives are to bring the message home to their supporters.

Finally, an even bigger problem emerged, a problem affecting not only ourselves but our entire world - environmental pollution. The next transition was on its way.

INTEGRAL PLANNING TO ENVIRONMENTAL PLANNING

One of the first cities to be confronted with the problem of environmental pollution by motor traffic was Los Angeles, where a special word was coined to describe this sometimes murderous mixture of smoke and fog - smog. Other cities had similar problems without the same density of traffic. In London, for instance, for many decades the death rate used to go up significantly with certain weather conditions when the world-famous London fog started to paralyse the life of the entire city. There, the pollution was the result of an excessive number of individual coal furnaces, and during recent years much success has been achieved in replacing coal by gas. The Ruhr Area in Germany, the industrial region of northern Italy and the Pittsburgh region in the United States are examples of other areas where environmental pollution soon became a severe nuisance.

In the early 1970s pollution was recognised as a worldwide problem and no longer a privilege of large conurbations. Human activities, for centuries trivial in their effects in relation to the size of our planet, had now grown to such dimensions that their damaging effects were comparable in scale to the global absorption capacities of nature. The findings of the Club of Rome, although in some respects perhaps over-dramatised, opened the eyes of many to the fact that in a closed system - and our isolated planet is such a system - no unlimited developments can be possible. There may be arguments on where the limits are, but not on the fact that they exist.

Now the responsibility of the planner grew even further. Not only was he responsible to his community, his region or maybe his country, but he now had to accept a responsibility in time as well. Measures proposed today and implemented tomorrow might seriously affect the lives of our descendants. Many things that can be done (and would have been done without qualms ten or twenty years ago) must now be reassessed from this perspective, while some other projects should be forced to implementation even though, in the circumstances of today, they may seem premature.

When environmental planning started ten or fifteen years ago, it was more or less restricted to traffic "cosmetics". Inadequate design was covered up by some face-lifting measures. Roads leading through built-up areas were replaced by by-passes (sometimes in the face of stern resistance from affected local merchants); noise barriers were erected along motorways; and the taxpayer was even prepared to support the preservation of some historic buildings which a few years ago would have been demolished. After all, the artificial environment is just as important as the natural environment, particularly in the cities, where there is little left of nature itself.

Now, however, environmental thinking should start entering the planning procedure in a much earlier phase. In order to be environmentally acceptable, a motorway may have to be designed to more far-reaching criteria than merely optimising earthworks or minimising costs. Noise will not be combated by double-glazing, but rather by tackling its causes. Air pollution will not be solved by prohibiting transportation, but rather by the development of new technologies. The depletion of our energy resources will not be reversed by demonstrations against nuclear power plants, but rather by giving some thought to a well-balanced modal split. Energy is anything but scarce in our universe. On the contrary, it is abundant; and where we should concentrate our efforts is in finding reasonable ways of using it.

In this respect, though, the latest transition is far from complete, let alone successfully so. Although important institutions such as the Environmental Protection Agency (EPA) in the US, the World Bank and others have recognised the importance of environmental impact statements, there is still a long way to go to convince politicians and voters alike that the cheapest alternative is not necessarily the best one. The energy crisis of 1973 has not yet produced any measurable effects on the developments of non-fuel-based technologies. On the contrary, the US is importing more fuel than ever before and has ruined its currency to a substantial degree through this laissez-faire policy. As late as in January 1978, the Swiss Federal Ministry of Customs and Finances refused to treat diesel fuel used for heat pumps in the same way as regular oil (with a reduced customs tax) because, as they put it, this was not an "effective contribution towards solving the heating problem".

CONCLUSION

As can be seen from these examples, the environmental planner has removed himself a considerable distance from the original traffic engineer. In the four transitions outlined above, he has moved from engineering to planning, from traffic to transportation, from a compartmented outlook to an integral approach, from local and immediate responsibility towards a global responsibility extending through time.

This changing role of the transportation engineer has not yet been fully recognised, either by our clients or by some of our colleagues. The narrow-minded outlook of many specialists is reflected in over-sensitive public attitudes which more and more frequently create serious problems in the dialogue between professionals and laymen. The credibility of our profession is at stake, and serious efforts to improve the conditions of our conurbations could be paralysed for many years. It is very important, therefore, that all concerned should accept their responsibility, and that those in charge of awarding planning contracts should become aware of the importance of their decisions. Unfortunately, the recession in the Western world has led many agencies to an anti-planning attitude. Nothing could be more wrong that that: one can afford not to plan only if there is an abundance of resources of every kind. Our present resources are not so abundant: we are restricted in terms of finance, energy and development space. It is among our gravest responsibilities towards society to point out how these limited resources can be used to the greatest benefit of all, and to see the problems of our own specialised tasks from a somewhat more elevated perspective.

The role of transportation planning in the management of the total urban system

4

Richard J. Brown   South Africa

SYNOPSIS

The transportation planning process has recently been the subject of universal criticism on the grounds that after considerable expenditure on major studies there has been little or no improvement in travelling conditions in the major metropolitan areas. Fundamental inadequacies in urban transport are often a direct reflection of shortcomings in the institutional framework within which urban transport operates, rather than a lack of specification in the planning models used in the process.

Some of the inadequacies and problems found in the urban transport management systems include the separation of integral management functions among various bodies with their own priorities and budgets. This has resulted in the planning process becoming an end in itself, rather than a means to an end, as the planning authority and the operating authority are often separate organisations.

The theory and management of systems is discussed, essential steps in the planning process are identified in the context of the urban system, and an organisational structure is proposed for effective improvements in urban travel conditions.

INTRODUCTION

Since the inception of the Urban Transportation Planning Process (UTPP) in the mid-1950s there has been a succession of papers and articles criticising the form and content of the process. Most of this criticism has come from within the body of people responsible for the process. In the first instance, the criticism dealt with the statistical performance of the models. Examples of such criticism are those by Kassoff and Deutschman[1] and Alonso[2] who reported in 1969. More recently, the in-house criticism has turned to the behavioural performance of the UTPP. Stopher and Meyburg, amongst many others, in both papers and books[3] [4] [5] have clearly set out the shortcomings of the conventional travel demand estimation procedures. Atkins[6] in his award-winning paper published early in 1977 concluded after a review of the state of the UTPP in the United Kingdom: "We have a series of excessively complicated and expensive models using unsubstantiated and biased techniques to provide information of dubious accuracy for answering the wrong questions."

In answer to much of this criticism it has been argued by Dial[7] that planners can plead that they have been misguided in their ignorance of the issues, and that they have lacked both the technical skills or the fiscal resources to plan for other than the automobile-dominated existence.

It is in this argument that lie the seeds of perhaps the biggest criticism of the UTPP. That is, it has been carried out in isolation of the management of both the transport systems and the total urban system. This breakdown is mainly due to the lack of a correct organisational structure for local government in general and urban transport in particular.

THE MANAGEMENT PROBLEM

The major problems in transport have shifted from the details of the planning process itself to those of the implementation of the solutions derived from the UTPP. In other words, the quality of transport will not be significantly improved until greater attention is paid to management elements, such as administrative, financial, legal and socio-political aspects of transport projects, than to the analytical elements, such as operational and economic justification.

Recently the Transportation Research Board Executive Committee,[8] at the request of the Commission on Socio-Technical Systems of the National Research Council of the US, prepared a list of the ten most critical issues in transportation. Included in the ten issues was one entitled "Intergovernmental responsibility for transportation systems".

The main points which the committee raised under this title were:

□ that the appropriate roles and functions of each governmental level in funding, construction management, regulation and control need to be examined;

□ that the planning and implementation of transportation programmes are still ill-defined and need to be clarified before there can be assurance that transportation policy and plans will be implemented;

□ that one of the most significant factors affecting the quality of area-wide transportation systems is the nature of the institutional arrangements for supplying transportation services;

□ that the uneven distribution of costs and benefits causes some essential institutions to be unwilling to participate;

□ and that the answer may be that areas require a specialist institution to coordinate and fill gaps in the broad spectrum of supplying transport.

Barratt and Le Maire[9] reported briefly on the problems that arose from the fragmentation of the various professional groups involved in the planning process in France at the end of the 1960s. This resulted in the introduction by the government of a system of planning on five levels which laid down the planning horizon, objectives, study content, approval procedures and financial and economic procedures. The new system was designed to bridge the gulf between the professional solution and wider national interests.

More recently, these problems have been supplanted by the introduction of political dogma into transport thinking. This is especially true in Britain, where in cities such as London and Nottingham highway-orientated transport plans have been superseded by those with a political bias. The world-wide publicity with which these plans were introduced was followed by analytical studies which have shown that the hopes and aspirations of the politicians were not met, and in the main the politically-based policies have been abandoned.

Gakenheimer[10] identified one of the major reasons for the demise of the purely professional approach as the fact that most of the actively pursued local issues in transportation planning are external to the movement system and therefore to the quantitative models which are central to the professionals' methodology. This leads to a confrontation between the broad travel requirements of a metropolitan area and the needs of the local communities within it.

The recent history of transportation planning in Boston provides a dramatic example of this quandary which has become a phenomenon throughout the Western world. Bellomo and others outlined in some detail the approach taken at a cost of $5 million (1970 prices) over a period of eighteen months to meet the shortcomings of the professionals' methodology in the areas of environmental impact, community disruption and socio-economics."[11]

A further major problem area concerns the coordination of the various operators. The cities of London and Newcastle provide good examples of how recent legislation in Britain has failed to provide a management framework within which the entire planning, implementing and operating process is undertaken by one authority.

The Transport (London) Act of 1969 contained three major provisions. It gave the Greater London Council control over London Transport; it gave the council increased powers in highway and traffic matters over the various borough councils; and it made it the responsibility of the council to prepare transport plans for all aspects of transport. The major difficulty in the present organisation framework, however, is that British Rail remains independent in the sense that, although it is represented on the major policy bodies, it has no obligation to adhere to the policies of these bodies if they appear to be against the interests of the railways. Before the implementation of this Act, the boroughs were in a similar position in relation to highway matters.

In the case of Newcastle, the implementation of the light-rail system has been the subject of a legal battle over operational rights between British Rail and the local public transport authority which was responsible for the planning and implementation of the system. Most of these difficulties have now been overcome, but only at the expense of costly delays.

The author[12] concluded after a review of newly implemented legislation in South Africa that "the proposed management structure envisaged in the Urban Transport Bill[13] does not appear to provide the necessary links between those most actively involved in the metropolitan transport industry - the politician, planner, operator or the public - and these links are essential if the decisions made are to lead to both more efficient and effective transport systems in any particular metropolitan area".

In summary it can be said that in many countries the transport planning process of the 1960s, largely carried out by professionals for consumption by other professionals, has been opened up by the inclusion of the politicians. Even though the ideal system has yet to evolve, it is likely that it will not be too long before the remaining institutional hurdles are overcome so that a system can be produced within which the management of the planning process will ensure the correct balance between the politicians as the executives, the professionals as the analysts, the community as the consumers and the operators as the managers. The remainder of this paper will be devoted to such an organisational system.

MANAGEMENT OBJECTIVES

As pointed out above, the present world-wide failures in urban transport systems are to a large extent the result of basic deficiencies in the management structure, allied with a dearth of people with management capability in what is essentially a public undertaking.

While the correct structure is not a guarantee that objectives will be attained, it is a prerequisite for an improvement in transport operating conditions in metropolitan areas.

Management as a science has been variously described. In the field of urban transport, the following definitions contains the essentials: Management is a discipline whereby the key resources of an undertaking are employed in such a way as to provide both short-term benefits and concurrently to ensure the longer-term growth and stability of the organisation.

From this definition certain key functions can be seen together as forming a logical decision-making process. They are assessing, planning, organising, implementing, evaluating, and revising.

As Shuldiner[14] states, the end value of planning lies in the decisions derived from the total process of which planning is an integral part. Unfortunately, urban transportation planning has not been implementation-orientated, but has been viewed rather as a process by which a "plan" is created. The plan is usually focused on the design of a horizon-year transportation network to the exclusion of other major aspects of urban transportation investment decisions. Because planning as a function becomes isolated from the central decision-making process, it often degenerates into little more than an elaborate, expensive and lengthy technical exercise in forecasting.

Johnson and others have warned of the dangers of specialised planning staffs: "All too frequently this staff assumes that its role is planning rather than facilitating the planning activities of line management. Left to its own discretion, this staff proceeds to set up goals and plans according to its own conception and premises and often develops elaborate research reports to substantiate its position."[15] Although this commentary is directed at large business organisations, its applicability to urban transport cannot be denied.

What constitutes a good decision-making process, and the part that planning plays in that process? Decision-making can be defined as the process of selecting from a number of alternative courses of action with a specific objective in mind. There are four phases involved in the process namely information gathering, conclusions, decisions, and execution.

If the process is to be effective, those who will be affected by the decision and involved in its execution, should be made aware of the issues requiring a decision and encouraged to participate in the data-gathering in accordance with their particular expertise. Those who have been involved in the data-gathering discuss and analyse the alternatives which are generated so that the conclusions can be a team effort. Decisions should be based on a rational choice of the alternatives and include the opportunity for feedback. In summary, decision-making should be decisive and systematic and should place the emphasis strongly on a diagnosis of the whole situation.

Within this process the first two phases are concerned with planning, which may be considered an activity in the present aimed at getting results in the future. The comprehensiveness of the planning function must be assessed in the following areas: time perspective, involvement, data collection and analysis, and objective setting.

Planning should be undertaken within a framework of time horizons in which projections and predictions are made on the basis of probabilities calculated from available data. Longer-term horizons are concerned with strategic planning, where data are global in nature. By contrast, for shorter-term planning the data need to be more specific. On the basis of the criterion of who can contribute, planning teams should be constituted to provide the manager with the necessary depth in problem understanding when he takes decisions.

The expertise of each member of each team is used to achieve a comprehensive programme of data collection and analysis. After the data have been analysed, objectives should be clearly formulated in terms of premises and values determined by top management. Where possible, objectives should be quantified and criteria determined.

It is within this framework that the proposed urban transport system decision-making process in general and the role of the transportation planning process in particular will be discussed.

A PROPOSED MANAGEMENT STRUCTURE

Cyert and March[16] viewed the management organisation as an adaptive social system with many interest groups integrated into a coalition. This system would develop mechanisms for avoiding uncertainty, engage in problemistic search, learn through experience and seek satisfying rather than optimal decisions.

When one attempts to put forward a management structure, it is necessary to study the functionl links between those actively involved in any transport system.

There are essentially four easily identifiable participants in the transport sub-system, namely the politician, the public, the planner, and the operator.

The politician. Public representatives are the top managers of the organisation. They have the necessary executive power, either by direct action or through the relevant legislative assembly. to affect meaningful changes in the environment within which the urban system functions. It is this important fact that the other participants appear to overlook when they attempt to make changes in the system. It is important that the politicians be actively involved in all aspects of the transport system so that their decision making may reflect an understanding of the functioning of that system.

The public. The public is most affected by the decisions made within the system, not only as consumers of the transport units supplied, but also in the wider connotation of the functioning of the urban system of which the transport sub-system is part. Thus, the needs and aspirations of the public, both as consumers and participants in the urban system, must be investigated so that the objectives fully reflect these needs.

The planner. The role of the planner must be seen as an extension of the manager's planning function, not as a replacement for that function. Those professionally involved in the planning function are concerned with the urban system as a whole. It is imperative that the transport sub-system at this level of planning be considered part of a much larger social, as opposed to technological, system. The professional planning teams should be multidisciplinary and should include economists, engineers and sociologists.

The operator. Those involved in operating the transport sub-system are in the main only concerned with the techno-economic aspects of this sub-system. This does not mean that their contribution should be restricted to implementing the decisions that are made. On the contrary, the operators must be involved at an early stage so that any objectives that are drawn up reflect the needs and capabilities of the transport industry. In this context, transport industry includes all those bodies involved both in providing the infrastructure and in operating the particular subsystem. Such bodies will include the road building authority, the traffic management authority and the public transport undertakings.

THE RELATIONSHIP BETWEEN THE PARTICIPANTS

The framework within which the relationship between the various participants is formulated can be represented by a tetrahedral form which provides four planes of activity made up of a series of six functional links. This framework is shown diagrammatically below:

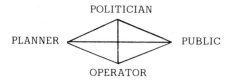

Prior to discussing the four activity planes, it is necessary to consider the two-way functions of each of the six links.

The links are considered in terms of the functional relationship between each pair of the four participants, thus resulting in six pairs.

Politician-Public. This relationship is basic to any Western democracy in that politicians make the decisions that affect the quality of the habitat within which we live, while the public as a body votes politicians into office and can remove those from office who make decisions which do not reflect the needs or aspirations of the public.

Politician-Planner. This link is representative of the employer-employee relationship, in which the politician directs the policy for the approach to planning while the planner balances the socio-economic planning variables so that the political body is fully aware of the consequences of its decisions.

Politician-Operator. This relationship, too, is basically one of employer-employee, in that the political authority directs policy and controls the implementation of the various transport undertakings, while the operator provides the operational and financial input and consequences on which decisions must be based.

Public-Planner. The relationship between the public and the planner is much less well defined but this should not detract from its importance. The role of the planner is to ensure that the public is made aware of the issues involved. In return, the public not only plays an important role in the formulation of system objectives but is also the source of much of the data on which the planning is based.

Public-Operator. The public is the consumer of transport units supplied by the operator. This is the basic relationship of economic theory.

Planner-Operator. This relationship is between two groups of professionals both of whom have clearly defined functions and thus depend on a high level of mutual cooperation.

The role of the planner is to supply a framework within which the transport sub-system can operate to obtain maximum benefit for the community as a whole. In return, the operator supplies inputs for objective setting and decision making in the form of data concerned with operational and financial aspects.

ACTIVITY PLANES
The four planes are in no way independent of each other, as each constitutes one phase of the decision-making process discussed earlier.

Public-Planner-Operator. At the base plane the public, planner and operator are concerned with information gathering, leading to the identification of the issues through the collection and analysis of data.

Politician-Public-Planner. The second phase of the process in which conclusions are drawn and overall objectives are set for the total urban system involves the politician, public and the planner.

Politician-Planner-Operator. The activity of the third phase of the process concerns the politicians and their professional advisers in taking decisions concerning the transport sub-system.

Politician-Public-Operator. The fourth and final phase involves the politician, the public and the suppliers of the transport units in the implementation of the decisions that emanate from the previous three phases.

ORGANISATIONAL STRUCTURE
Having established the role of each of the four participants, the relationship between each pair, and the involvement of each of the participants in the four phases of the decision-making process, one must translate this framework into an organisational structure which will ensure that the links representing the six relationships and the four activity planes are fully integrated.

Of prime importance in any organisational structure for a public body is a high degree of political accountability. The transport-orientated decisions that are made generally affect only one particular locality; thus, politicians of that area should be responsible for those decisions. If all decisions are made at the national or state or provincial level, important local issues become insignificant compared with national issues such as defence, internal security and economic policy, and public dissatisfaction with local transport policy and implementation is not reflected through the ballot box. However, a degree of involvement at the national level is still required.

Two prime functions have to be performed at the national level. The first is that the financial needs of metropolitan transport are fully represented when the national budget is prepared, and the second is that broad policy decisions have to be made in areas where the needs of metropolitan transport and national requirements interface. As part of this latter function, research should be coordinated and advisory services rendered at the local level so that limited expertise in this field can be used effectively.

According to Dial,[7] one of the lessons to be learned from the present history of urban transportation planning is that the planning, implementation and operation must be coordinated without an artificial administrative and jurisdictional partitioning of functions and resposibilities.

While no detailed discussion is possible here of the advantages and disadvantages of forming metropolitan authorities with direct links to the national level rather than to provincial or state authorities, a few comments may be ventured. The public representatives and through them the public of a metropolitan area should be wholly involved in the decisions affecting that area. In many parts of the world provincial authorities tend to favour the needs of the rural rather than the urban communities over which they have jurisdiction. In order to give those affected by the decisions a meaningful say and to ensure that those who make the decisions are politically accountable, the organisational structure should be based on a metropolitan authority directly responsible to the national level. The kind of authority envisaged would be similar to the Greater London Council in that all matters pertaining to the metropolitan area, including transport, would be dealt with by the metropolitan authority, while other local matters would be handled by the municipalites and peri-urban authorities within the metropolitan area.

Within such an authority there should be three divisions under a director of metropolitan services which would be responsible for the various aspects of metropolitan transport affairs. Each division would be responsible through the director to a sub-committee of the proposed metropolitan council. The three sub-committees would be responsible for metropolitan planning, transport operations, and advisory services.

Metropolitan planning. This division would be responsible for carrying out the overall physical and economic planning of the metropolitan area at the strategic level. Through this division, data would be analysed and the many variables balanced to produce the basis for decisions. Departmental heads would be appointed for socio-economic planning, land-use planning, transportation planning, and a data bank.

Transport operations. The implementation of the decisions taken at the end of the strategic planning phase would fall to the division of transport operations which would consist of the following departments: operations, train transport, bus transport, road construction and maintenance, road traffic and pedestrian management, coordination and short-term planning, and finance.

Advisory services. Under the subcommittee responsible for advisory services, the division would be concerned with liaison between the authority and the public. The division would consist of the following departments: information, environment and safety, data acquisition, and consumer affairs.

Figure 1 (page 38) shows the intended organisation structure within the proposed metropolitan authority. Of course, in many countries the expertise or manpower is not available to set up this kind of organisation for each large urban area. Nevertheless the need for such an organisational structure will become increasingly more obvious and thus we should start to work towards it. Two aspects of this preparation are immediately obvious. First, we should step up the resources available for the education and training of managers within the various transport orientated disciplines. Second, we should seek a better balance between the public and private sectors in the transport fields, so that the private sector is used as a source of specialist knowledge rather than as a source of manpower.

CONCLUSIONS

Much of the recent legislation on urban transport around the world will undoubtedly assist in alleviating the short-term problems, Unfortunately, in the field of management many of the existing and proposed organisational structures neither eliminate existing management problems nor provide the vehicle for effective decision making.

More specifically, transportation planning must be subdivided into those aspects which are of a strategic nature, and those which concern matters of coordination and short-term budgeting. At the strategic level, the planning must form an integral part of the total urban planning process while the medium (coordination) and short-term planning functions should be divisionally (transport services) and departmentally (each mode) orientated. Planning at all levels must be considered an aid to the management decision-making process and not as a self-fulfilling exercise.

While the required manpower is admittedly not available to set up the suggested alternative organisation, it is considered that the need for such an organisation will become increasingly more obvious with the passage of time. In the meantime, efforts should be made to work towards this solution by training manpower and setting into motion the necessary investigations prior to making major changes in the organisation of the governing authorities for metropolitan areas.

REFERENCES

1   Kassoff, H., and Deutschman, H.O., Trip generation: A Critical Appraisal (1969), Highway Research Record 297.
2   Alonso, W., The quality of data and choice and design of predictive models (1968), Highway Research Board special report 97.
3   Stopher, P.R., and Meyburg, A.H., Travel demand estimation: a new prescription (1974), Traffic Engineering and Central.
4   Stopher, P.R., and Meyburg, A.H., Urban transportation modeling and planning (1975), Lexington Books.
5   Stopher, P.R., and Meyburg, A.H., Behavioural travel demand models (1976), Lexington Books.
6   Atkins, S.T., Transportation planning: is there a road ahead? (1977), Traffic Engineering and Central.
7   Dial, R.B., Urban transportation planning system: philosophy and function (1976), Transportation Research Record 583.
8   Transportation Research Board, The ten most critical issues in transportation (1976), Transportation Research News.
9   Barrett, R., and Le Maire, D., Urban transport planning: finding the right level (1973), Traffic Engineering and Central.
10   Gakenheimer, R.A., The transition in urban transportation planning (1973). High Speed Ground Transportation Journal, Vol 7, No 1.
11   Bellomo, J., and Lockwood, C., Metropolitan transportation and land-use planning: an evaluation (1973), Transportation and Environment: Policies, plans and practices. University of Southampton.
12   Brown, R.J., An alternative organisational structure for managing the Metropolitan Transport System (1976), SAICE Blydepoort Transportation Conference.
13   Republic of South Africa, Urban Transport Bill 1976.
14   Shuldiner, P.W., The federal role in urban transportation planning (1973), Transportation and Environment: Policies, plans and practices. University of Southampton.
15   Johnson, R.A., Kast, F.E., and Rosenzweig, J.E., The theory and management of systems (1975), McGraw-Hill.
16   Cyert, R.M., and March, J.G., The behavioural theory of the firm (1963), Prentice-Hall.

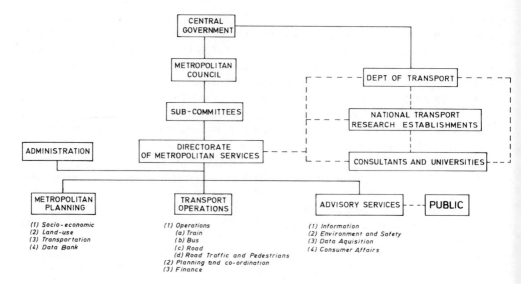

Figure 1. Suggested organisational structure for the transport aspects of a metropolitan authority

Using urban planning concepts 5
to reduce travel and
improve the environment

Karl Krell   Germany

SYNOPSIS

Many environmental problems in urban areas arise from commuter traffic with private cars, owing to the concentration of work places and stores in the city centre and residential areas far away from the city. While improved organisation of commuter traffic would be helpful, it would be better to avoid commuter traffic by motor vehicles with a new town concept: rejoining work and living places. If work places (light industries, administration, banks, etc), stores and parking areas were constructed in a ring around living areas (600 m x 600 m), many people could walk to their working areas and stores. The advantages would be energy savings, reduction of noise and exhaust gas emissions, less visual intrusion, and protection of residential areas against existing noise and air pollution by an appropriately planned ring-shaped design of closely structured surrounding buildings.

ENVIRONMENTAL PROBLEMS CAUSED BY THE URBAN TRAFFIC SITUATION

Most environmental problems in larger cities are caused by the fact that work places and dwellings are separated by a great distance, and by the growing concentration of work places and department stores in blocks of buildings in the centre of cities.

The assumption that a separation between work place and dwelling was appropriate was certainly not without foundation in the 1920s because at that time most work places with their noise and pollutants had a considerably detrimental effect. Nowadays work places and stores hardly disturb neighbouring dwellings.

When workers relied on public transportation to take them to work, the concentrated blocks of work places in the city could still be accepted. With the increase in space taken up by work places in the city, however, more and more workers had to travel to their employment.

The quality of the urban environment took a negative turn when more and more workers became able to run their own cars. For those owning a car and paying the fixed costs anyway, it is cheaper, more comfortable (despite congestion) and quicker to use it to the city centre than public transport would be. The consequence of the conditions just described was a steady growth in the number of private car commuters, which in turn led to a continuing increase in undesirable side-effects on the environment. The symptoms are the following.

□ A considerable amount of energy (fuel, electricity) is consumed on the journeys to work, and in the provision of durable goods, which are becoming increasingly scarce and expensive. (In West Germany, the journey-to-work trips account for more than one-third of all vehicle miles.)

□ Congested streets in the city during rush hours lead to increases in energy consumption, emission of pollutants and delays.

□ More and more space is used for new road structures and taken away from food production or recreation uses, because journey-to-work trips require substantial pavement space (all commuters need to be on the road at the same time) and twice the parking lot area (at home and at work or at park-and-ride stations).

□ The street loses its significance as a playground for children and place of communication for pedestrians. Wide and densely trafficked streets cause severance effects (separation of communities).

□ Street safety is jeopardised by drivers. The consequences of accidents are worst for unprotected road users (pedestrians, two-wheelers).

□ With the growing awareness of the environment, traffic noise is increasingly felt to be a nuisance, and also disturbing life in roadside houses.

□ New streets and parked cars everywhere cause visual intrusion. The historical character of towns is detrimentally affected and sometimes even destroyed.

□ Taxpayers in high income brackets leave cities and settle in the environs. The infrastructure in the city is no longer fully used while real estate in the suburbs is denied other uses. Funds are needed for new infrastructures.

PREVIOUS ATTEMPTS TO IMPROVE THE ENVIRONMENTAL SITUATION IN CITIES

In most cities in industrialised nations attempts are being made to treat the symptoms mentioned.[1] Despite the activities of certain political groups pleading for the abolition of the car, the right to private car ownership is hardly questioned in the official politics of any nation. Yet, most attempts at improving the environmental situation are aimed at relieving the undesirable side-effects of motorisation.

In order to save energy and reduce the emission of pollutants, the reduction of petrol and oil consumption and the emission of pollutants per kilometre driven is made a condition for car manufacturers.[2] In West Germany, for example, the future emission of passenger cars will be one-tenth that of 1969. Entirely different engines and a substitute for petrol and oil can only be expected after the turn of the century.

An attempt is made to reduce congestion by means of signalling on new roads or lanes. New and wider roads, in turn, cause environmental problems as they require more land, impair the townscape and cause severance effects. The easier access to a city is made, the bigger the parking problems become. If the concentration of work places in the city is high, no town planning concept can possibly provide the streets and parking lots necessary to enable each individual to use his own car and leave it in the city while he works.

To make better use of the space provided by pavements, more and more reversible lanes are tested. For example, two-thirds of a road's lanes are available to city-bound traffic between 2 a.m. and 2 p.m., and to out-bound traffic for the rest of the time. Such solutions, however, do not reduce the parking problem.

To improve traffic safety and reduce congestion and severance effects, increasing use is made of signalling systems.[3] Above all green waves are used in an attempt to maintain the traffic flow. However, these efforts are jeopardised by an unfavourable pattern of road networks and by unrealistically high demands. Some cities, such as Berne, therefore used the so-called "gatekeeper" or "control of access" method to shift unavoidable congestion in the city to areas where congestion would not affect roadside houses to the same extent. But this only shifts the disadvantages for traffic from one area to another. Signal-controlled pedestrian crossings have been established to improve safety and, to some extent, reduce severance effects.

Streets, particularly shopping streets, are closed to traffic to give more rights to pedestrians. But the improved environmental conditions of shopping streets have hardly had a chance to benefit the quality of housing. Hardly anyone lives there any longer. In West Germany, an action called "restricted traffic in residential areas" was started to move out-of-town traffic (driving or stationary cars) out of these areas (cars belonging to neither the residents of the area nor their visitors). For this purpose "residential traffic cells" have been established where through roads are cut off and traffic may not move faster than 30 km/h. In these cases, the removal of the constructional separation between carriage-way and sidewalk has also been considered. The expectation that the noise level would be reduced at a speed limit of 30 km/h was not realised, owing to persistence of driving habits. However, it is expected that the measure will lead to improved safety and reduction of the visual intrusion caused by cars.

Efforts are being made to reduce noise emission as much as possible by improvements to the motor vehicle. Obstacles are tyre noise and the "noisy" driving style of part of the driving population. Technical measures alone will therefore not reduce traffic noise to a point where disturbance drops to zero.

Town planners are thus left with the task of reducing traffic noise levels by the following means:[4] increasing the distance between pavements and dwellings, screening, constructing highway tunnels, and installing soundproofed windows.

In urban areas, substantial distances between pavements and apartment buildings or homes are usually not possible. In general, there is no space for constructing earth-banks as a noise abatement measure either. Nor would noise screeens or earth-banks fit into the townscape, except in special cases. In addition, the noise level in front of the highest and closest dwellings is only reduced by about 10 dB(A) through earth banks. The construction of roads in tunnels is generally not possible because of the high cost and power demands for illumination and ventilation.[5] The screening effect of cuts, not requiring artificial lighting and ventilation, is often over-estimated. Nor are they a sound solution, owing to their severance effects. Elevated highways with lateral screens are fairly effective acoustically and their severance effect is less, but they do not always fit into the townscape. Even if highways were to be placed on roofs, the ramp connections needed to link them with the remaining network would raise problems of aesthetics.

Soundproofed windows can effect a noise reduction of 40 dB(A) and more, but they do not protect gardens, terraces, balconies and loggias. In West Germany many cities grant subsidies for soundproofed windows (generally 50 per cent) in order to prevent taxpayers in high income brackets moving out of the city to settle in the suburbs. These measures have had unexpected effects, however. The better soundproofed windows are in reducing traffic noise, the more noise-aware citizens become angry at the now audible noise of their neighbours.

These measures do effect certain improvements but do not bring about significant changes to arise from measures which noticeably reduce the number of cars used for daily journey-to-work trips into town, however. Politicians, who usually support a free choice of transportation, also favour the idea of restraining the use of the private car for journey-to-work trips. So far, there are two approaches: attempts to induce car owners not to use their car on journeys to work by offering them "attractive" public transport systems, and attempts to increase the number of occupants per car on journeys to work.

Both solutions involve difficulties, however. Many citizens with one or more cars have chosen the dwelling which affords them the highest degree of freedom - the single-family house - without regard to public transport and the location of stopping places. An attractive and economic public transport service for residential areas of single-family houses cannot be achieved. Workers are therefore obliged to use their cars at least to reach a convenient stopping place (park-and-ride, kiss-and-ride).

The supporters of subway and bus systems have recognised that public transport cannot be made as attractive as would be desirable. The many stops necessary for getting on and off and to fill vehicles do not permit high journey speeds. Transfers and long distances to and from stops also increase journey times. On average, users of public transport facilities require 1.5 to three times the journey times needed by car users. An attempt has been made to raise the journey speed of subways by faster starting and braking manoeuvres, but this requires additional energy.

On subway systems in West Germany, generally fully occupied only during the morning and afternoon rush hours, energy consumption per passenger-kilometre is, at best, one-third less than that of a private car with 1.3 occupants. It was not surprising therefore when a study[6] pointed out that in West Germany the energy consumption of private car traffic would only decline by about 12 per cent if 60 per cent of drivers using their own cars to work were to use subway systems, which would still have to be provided! If planned energy-saving improvements in car design are successful, a passenger-kilometre by car, even at the present low car occupancy, might not require more primary energy than a subway system. The question whether noticeable safety gains can be achieved by the restraint of private car traffic still needs to be settled.

Subway costs per passenger-kilometre are generally higher than car operating costs in a restricted sense (fuel, tyres, etc). A person who owns a car and is not willing to do without it will generally find it cheaper, under conditions of free competition, to use his own car than to use public transport facilities. Public transport is therefore subsidised by many cities, but the required subsidies are so high that further increases would be beyond the reach of public funds.

On balance, buses are a better solution. In West Germany, buses require only one-third of the energy required by private cars (per passenger-kilometre). To improve their attractiveness, efforts are being made to increase journey speeds by means of priority lanes and signalling.

Despite the considerable investments made in public transport facilities, these have not become attractive enough to induce large numbers of private car owners to use them instead of their cars. Therefore, it is being deliberated whether coercive measures could increase the use of public transport facilities. No free country has yet found a satisfactory answer to the question of who should be forced to use a different means of transport, however. A solution by means of cost, e.g. road pricing and higher parking rates, is generally considered an anti-social measure. A prohibition of permanent parking (more than 3.5 hours) in the city for non-residents might generally be considered acceptable.[1]

According to the study[6] mentioned before, about 10 per cent of the energy consumed by private transport in West Germany could be saved if it were possible to increase the car occupancy of commuter traffic to 3.3 passengers per car. It will obviously require special incentives to induce individuals to use car-pools. On roads leading to the city, it has been found necessary not to levy charges on fully occupied cars, but rather to allow these to use special lanes in order to save the waiting times on the approach to toll gates.

THOUGHTS ON REMOVING THE MOST IMPORTANT CAUSES

If there is no basic intention of preventing private car ownership, care must be taken that the need for undesirable car traffic does not develop. (In any case, this is better than to allow needs to develop and then make their satisfaction difficult or impossible.)

As already mentioned, the daily journey-to-work trips are considered undesirable since they add much to congestion. Since work places in the city (banks, insurance companies, offices, mechanical workshops, department stores, etc) nowadays do not necessarily impair dwellings, the most obvious solution would be to establish dwellings at walking distances from the work place. If private cars were no longer used for the daily journey-to-work trips, the motorised traffic in cities could be reduced by 30 to 60 per cent and there would also be considerable reductions in energy demands, emission of pollutants and noise as well as the space required for streets.

It is clear that the reuniting of work places and dwellings and the reductions mentioned cannot be achieved within a short time. In the existing historical towns, such changes probably cannot even be accomplised in any consistent manner. In development and redevelopment areas, however, such a concept could be tried. Intermediate solutions may already be a help.

It needs to be mentioned, however, that real estate prices and tenancy laws which discourage moving often work against such concepts.

The ideal solution would be buildings which accommodate work places and parking garages and, at the same time, function as noise screens for the dwellings placed behind them. Figure 1 shows such a solution.

An important requirement for reuniting work places and dwellings in cities is the decentralisation of work places and stores. The Brazilian city of Curitiba with a population of more than a million has made provision for this solution in its planning policy. Work places and stores are to be established in a radial pattern on both sides of the large through roads to the centre of the city. As the town grows, these establishments will also grow and expand in an outward direction. The work places are built so that they simultaneously protect the residential areas behind them against traffic noise and air pollution. The through roads accommodate special bus priority lanes, lanes for public transport, through traffic and service lanes.

The most logical urban planning solution would be a compact annular development around a residential area with four- to six-storied buildings housing work places, stores and parking lots. If the pattern of the network of through roads is regular, a unit of dwellings and work places could have a size of 600 x 600 m (Figure 2). A unit of the same size would be possible in a radial road network. A unit of this size would guarantee sufficient noise protection against outside traffic noise and could accommodate work places, quiet apartments and parking places for about 5,000 people.

In each case, however, a decision is needed on the cars to be allowed into the building unit: all tenants, or only cabs, vehicles of the handicapped and emergency services? Pedestrians should always have the right-of-way in the interior of the unit. The size of the unit should be such that work places and stores in front of the building units, as well as bus stops outside the units, can be reached on foot and distances are no longer than 600 m. If a system of such units is linked to a subway system, the most logical place for subway stops would be the centres of the units.

The suggested solution of unit development may at first raise some doubts on how it should be fitted into urban planning concepts. If building fronts are designed to blend with the townscape, annular development accommodating work places, stores and multi-storey car parks should not look worse than the familiar ribbon development in large cities. Indeed, unit developments may provide better conditions for sound urban life than purely residential areas.

REFERENCES

[1] Krell, K., Gedanken zur Einschränkung der Kraftfahrzeugbenutzung (1974) "Der Städtetag" H. 4, S. pages 217-24.
[2] Krell, K., Verminderung der Verkehrsimmissionen (1977) "Strasse und Autobahn" H. 9, S. pages 357-64.
[3] Krell, K., Probleme der Sicherheitstechnik im Strassenwesen (1974), "Zeitschrift für Verkehrissicherheit" H. 1, S. pages 21-45.
[4] Krell, K., Techniken und Probleme der Abschirmung von Strassenverkehrslärm (1977) "Strasse und Autobahn" H. 4, S. pages 180-7.
[5] Krell, K., Sollen neue Strassen in Tunnels geführt werden? (1978), "Beratende Ingenieure" H. 7
[6] Schwanhäuser, Ermittling des Energieverbrauchs und der Schadstoffemissionen im bodengebundenen Personenverkehr bei alternativen Verkehrssystemstrukturen im Jahre 1985 (1978), Forschungsbericht für den Bundesminister für Verkehr (noch unveröffentlaicht)

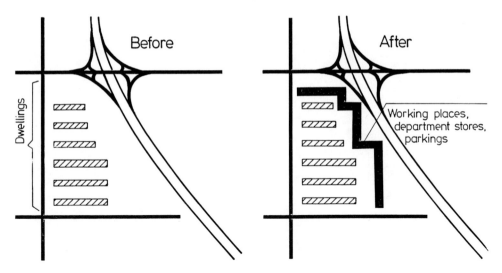

Figure 1. Protection of apartment houses against traffic noise by means of a building accommodating offices

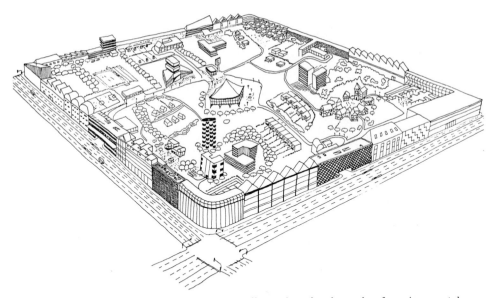

Figure 2. Ideas on a work-place and home cell meeting the demands of environmental protection

Social impact and transport technology: some policy considerations

6

E. A. Rose and P. Truelove   United Kingdom

SYNOPSIS

The question of how far transportation investment decisions can and should be informed by social considerations provides the context for an examination of specific examples of rapid transit projects in Britain and France. The rationales which inform investment decisions are compared, and ways of broadening the basis for evaluating policy options are considered.

Although institutional differences, variations in the level of urbanisation, and factors such as the timing of decisions, all influenced the eventual outcome, there were a number of similar factors which played an important, if not decisive, role in the decision-making process and its characteristic products.

In every case, the influence of technocrats, either at central or local government level, was a dominant factor in the genesis of proposals, and in no case was the influence of the output from mathematical transport models decisive. The development potential of new or upgraded rail-based public transport was often a factor in decisions, but usually in the context of city centre redevelopment or other suburban growth, rather than in the context of policies for renewal and revitalising inner urban areas.

The evidence derived from the case studies suggests that proposals for new and improved forms of public transport in the form of rail-based schemes to secure the policy objectives of economic and social regeneration in inner city areas, are not likely to be cost effective. Other more modest transport proposals relying on a less comprehensive but more diverse bundle of policies are likely to respond better to problems and permit social and economic objectives to be pursued.

INTRODUCTION

When new public transport facilities are planned in different countries, it might be expected that while different factors would influence the process whereby a system is selected and evaluated, the underlying conditions would be similar in some important respects. Urbanisation and increasing traffic congestion are well-known and world-wide phenomena. In practice, however, the evidence points to the diversity of metropolitan and urban situations and the relative paucity of technological/policy responses. There is a notable absence of requisite variety.

The conventional and simplistic explanation for the 1960s boom in new rapid transit construction held that it stemmed from "the realisation that only reserved track systems (could) offer an attractive alternative to the private car and buses or trams mixed with other traffic in heavily congested road".[1]

In fact, the explanation is far more complex. It has been suggested that seldom has a cause been so fervently and unanimously championed as that for rail rapid transit, the advocates of which constituted, in the United States at least, a coalition that included the whole spectrum of interests from rabid environmentalists to downtown bankers. "Central-city bound suburban commuters who favour sprawl unite with frustrated planners anxious to recentralise the urban map. Inner city civic groups, central city mayors, and real estate investors are at one with the news media and the roving band of rail-equipment producers and consultants. Such unity before what is a complex urban problem cannot but arouse suspicion."[2] Detailed case studies relating only to the United States are presented in evidence.

A parallel and somewhat contradictory view holds that urban road building has been the result of a coalition of interests between the state and the car industry. Dupuy has pointed out that forecasts of the production of cars are given factors and inputs in the transport modelling process.[3] The provision, or proposed provision, of road space is justified by, and in turn justifies, the forecast production of motor vehicles. While modelling techniques originated in the US, Dupuy suggests that conditions in the 1950s in the US, and in the 1960s in France were analogous in certain important respects. In both cases, the role played by the motor car industry in the development of the national economy was very important, and in both cases led to massive intervention by the state in the building of roads.

These apparently different interpretations of urban transport policy formation may well both be right; the two sides of the same coin. But both examples serve to demonstrate how far removed real transport policy decisions and practice have been from the so-called detached scientific approach implied by the use of transport models in pursuit of "rationality" in transport decision making. It is both illuminating and important, therefore, to identify and describe the decision processes, and the unanticipated as well as anticipated consequences of both policies and procedures.

COMPARATIVE POLICY CONTEXTS

The rationales for the new construction work in progress in the UK, notably upgrading the extensions to the existing and modest suburban railway systems in Glasgow, Liverpool and Newcastle, may be found in decisions made as a result of major land-use transportation studies.[1]

By way of contrast, the decision to extend a Paris metro line to Creteil to the east appears simply to have flowed from the political decision to embark upon the major peripheral town development at Creteil, which has become in effect a major new residential centre or growth point; a new town in all but name, with a population of approximately 50,000.

In questioning whether the difference in approach is more apparent than real, it is necessary to consider the context within which the decisions were made.

DIFFERING RATES AND LEVELS OF URBANISATION

Although the UK and France are both half way to a fully motorised society,[4] and most families may soon have at least one car, the more important indicator for comparison of policy and practice is probably the rate and level of industrial urbanisation. There are important differences to be noted here. Britain could be said to have been predominantly urban and industrialised by the end of the nineteenth century. By contract, France reached a comparable stage only recently.

Intra-regional population change in the UK rather than inter-regional population growth has been the decisive factor in bringing about a far-reaching reappraisal of the validity of the Barlow Report's recommendations, particularly as they affected London and perhaps to a lesser extent Birmingham. But the central fact is that conurbations as well as central cities have been losing populations and manufacturing employment at an alarming rate since the beginning of this decade and the post-war consensus embodied in the Abercrombie Greater London Plan no longer exists. If transport in London is to be improved, it is certainly not to meet increased population pressures within Greater London boundaries. National population growth has been distributed and redistributed beyond the green belt and in the wider areas of South-East England. The new towns have played a modest role as well.

The picture was very different in the Paris region, as reference to the wider planning strategy for the Seine Basin makes abundantly clear. Thus, the decision to extend the metro line to Creteil may be directly related to the rapid growth of population in the Paris region in the 1960s. Strictly speaking, Creteil was not and is not a new town and its own characteristic growth reflects the magnetic influence which Paris continued to exert on the rest of France throughout the period under discussion.

There is no comparable context for recent UK decisions. New investment in rapid transit has not been allied to developmental objectives, such as suburban extensions (the extenstion of the Piccadilly tube line to Heathrow airport is hardly relevant in this context). The prime objective of policy has been to upgrade the existing rail systems and improve levels of service and interchange. The emphasis has been on central area improvements and in particular the linking of formerly separated terminals.

This policy is exemplified by the Reseau Express Regionale (RER) in Paris, and the abortive "crossrail" proposal for London. This provided for new tunnels beneath central London, built to British Rail rather than underground dimensions, to be used by the suburban railway system. In Liverpool, a single track loop line, completed in 1977, now distributes railway passengers through the central areas, and provides connections with a new underground link between suburban lines formerly serving separate terminals. Interestingly, these works resemble proposals first made by the railway company in 1925-6, at a time when rapid urbanisation was still in progress in the Wirral, partly helped and influenced by the rail and road tunnel links across the Mersey estuary, improving accessibility to the centre of Liverpool. It may be a case of too little, too late in the case of Liverpool, where the central planning problem is arguably to prevent both people and jobs from leaving the city.

TECHNICAL STUDIES, TIME HORIZONS AND UNCERTAINTY

Two necessary if not sufficient conditions for the formulation and implementation of rapid transit proposals are the availability of adequate resources and the political will. The implementation stage in the decision process is often the missing link, however, as many policy studies have shown. Moreover, political will expressed at the local level may require major support in more tangible resource terms from the Treasury in Whitehall. Political good will is not enough. The lessons to be drawn apply not merely to transport proposals. The policy environment and indeed the economic climate may change rapidly in conditions of uncertainty.

The time horizons embraced by the "one shot" nature of transport studies and the time scale of the physical implementation and planning of proposals become progressively more difficult to reconcile with changing ideas of what process and adaptive planning should seek to do. But perhaps the truth of the matter was that urban surgery on the grand scale was politically and economically unacceptable in urban contexts which were already declining; unacceptable to government, unacceptable to real estate interests, and unacceptable to various local communities that believed they would be adversely affected by the change of activity patterns that such schemes might bring about. Exceptions prove the rule, and certainly the schemes that did go ahead were modest by French standards and were regarded as environmentally acceptable and economically beneficial.

Therefore, in the course of the 1960s and 1970s, a number of English cities considered the possibility of constructing new rail-based transit systems. The timing of these studies seems to have been quite as important, so far as decisions were concerned, as the technical content and quality of the studies. Political and economic factors predictably weighed heavily in the case of UK studies. Thus, the decision to go ahead with a light railway metro on Tyneside in North-East England was made at a time of government concern with rising unemployment in this part of the country, and at a time when central government had resources available for new infrastructure. It is probably reasonable to suppose that the view of central government was that such a scheme would assist the region as a whole. But no sophisticated economic or social evaluation was carried out.

Similarly, the precise influence which land-use transportation studies have had in practice is hard to determine and awaits more definitive treatment. The rise and fall of the London Transport Study is well documented. But the various proposals were weakly linked to rail improvements for reasons which relate to historical and institutional relationships between transportation operators.

The evaluation of the proposed central area tunnel across Manchester in the SELNEC Transportation Study (1972) was somewhat inconclusive. In other cities, the major land-use transportation studies were often too late, despite their cost, to be a crucial factor in the decision process. In an important sense, they were one of the first types of policy informing/planning exercises to be overtaken by events - the mounting demographic, economic, behavioural and political turbulence since the late 1960s. For example, the MALTS Transportation Study for Merseyside reported in 1969. The Act of Parliament authorising the construction of the new loop railway under the city centre was passed in 1968. Interestingly, the earlier, more naive traffic studies of about 1960 may have been more influential, but in an unexpected way. These demand studies forecast an enormous growth in car usage, often predicting for 1980 traffic flows that now seem unlikely to arise before 1990 or 2000, if at all. The effect of these forecasts was both to encourage local authorities to embark on massive road construction programmes, and to look for new ways of meeting demand for movement. e.g. by developing the railway system.

At the time these forecasts were appearing, two more important factors were at work. City planning departments were being established within the conurbations, and new rapid transit technologies were becoming available. Thus, in Liverpool, the loop railway proposals were first revived in the draft city centre plan of the first city planning officer, Walter Bor, before consultants were engaged to produce the final plan. In the case of Manchester, the catalyst for study of rapid transit was the initiative of a firm of civil engineering and building contractors, Taylor Woodrow, who held the concession for the Safege monorail, a French system using rubber-tyred wheels supported by an overhead beam.

Obviously, the approach of a local authority to a new form of public transport would be extremely complex and influenced by the organisation and personalities within the local government system, but what seems absent is any clear policy guidance from central government or indeed direct influence in the genesis of specific proposals. Nevertheless, it must be pointed out that the then Ministry of Transport encouraged studies into new rapid transit developments by making 50 per cent grants towards the cost of studies, and its divisional officers were represented on the steering groups of all transportation studies. So just how did the type of government administration influence the nature of the transport decisions made?

THE ADMINISTRATION OF TRANSPORT POLICY
It might be thought surprising that close parallels may be drawn between transport policy making and the history of nations. In their study of transport in relation to politics and public policy Thoenig and Despicht make this trenchant observation: "The field of transport policy is, in fact, extremely well organised with strong tendencies towards technocracy."[5] They point out that France has had a centralised national state authority which has concerned itself with the regulation and planning of transport at least since Louis XIV.

It was not an accident that the Corps des Ingenieurs des Ponts et Chaussées was a seventeenth century creation, institutionalised in 1740. In France, with or without planning, transport is a sector over which state and officials exercised and still exercise a strong hold. Thoenig considers that this bureaucratic interventionism is justified by a long tradition which regards transport as one of the special instruments of state action aimed at ensuring at one and the same time economic development, control of the country from Paris, and military defence.

There is no doubt about the power of the Ponts et Chaussées service. Thoenig refers to a pressure group of Ponts et Chaussées engineers at the top of the government machine ensuring the permanence of railway policy. Dupuy refers to transport models as an instrument of the power of Ponts et Chaussées engineers in furthering road building.[3] Transport models present a technique which is comprehensive to engineers and provides the "scientific" backing to their proposals. Indeed, if similar transport models were applied throughout the nation, one might expect the resultant policies to conform to a national pattern.

While the UK lacks a long history of state intervention in transport (other than the safety and anti-monopoly legislation concerning Victorian railways) there is a much longer history of land-use planning. It was not until the 1968 Planning Act, however, that the idea of integrating land-use and transport planning was written into statute, although the influence of city planning officers (who in many cases were previously or remained highway engineers) on urban transport policy was beginning to emerge by the early 1960s, i.e. in time to contribute to decisions on, for example, long-cherished proposals for inner ring and radial roads, as well as urban railway development. As in the case of the Liverpool loop railway, some of these road proposals had existed for a long time. For example, Birmingham's inner ring road was first proposed before the second world war. However, this influence was exercised informally in, for example, the preparation of non-statutory city centre plans, and not through any specific transport responsibilites. Indeed, local government structure throughout the 1960s continued in a form unsuitable for the requirements of modern transport policy.[6]

The 1968 Transport Act which contained positive measures for the development of public transport was not operative until 1969, and agreement was not reached before 1973 between the various bodies with some responsibilities for transport planning. While the 1968 Act marks the first recognition of government of the social and environmental aspects of transport policy, the organisation of the Ministry of Transport indicates that its traditional functions, such as the trunk road programme, occupy by far the majority of the civil servants working for the Ministry. Indeed, functions added more recently appear to work in parallel with established functions rather than in any rational hierachy. Thus, there is an under-secretary for transport policy, which embraces a unit charged with the supervision of transportation studies, and, for example, an under-secretary holding a post of equal status in the hierarchy with responsibilities for highways and bridges.

However, an important difference from the French situation lies in the educational background of the senior civil servants. The English tradition of the non-specialist administrator contrasts strongly with the engineering-based technocracy of French administration. In describing the attempts in the Fourth and Fifth plan to link transport with the planning process, Thoenig comments: "Between the planners and the officials of the traditional ministries, an authentic planning network was created, over and above the hierarchial pyramids and divisions. Inter-personal communication was facilitated by the fact that the majority of the members of the network belonged to the same corps (Ponts et Chaussées)."[5] This network within central government does not readily permit innovations in transport policy to emerge within the local government bureaucracy.

This perhaps marks the essential difference between the French and the English approach to decision making on the development of rapid transit. The apparently rational use of transportation studies in aiding policy decisions has been used as something of a smokescreen, for in many cases the study was used to justify a decision that had already been taken. The difference between the French and the English approaches lies in the apparently greater scope for transport policy innovations that rests with the technocrats employed within local government.

Whether tentative innovations emerging from city authorities could flourish without at least tacit central government approval is another question. When it comes to financing new transport investments, British central government maintains as effective and pervasive a control as does the French government.

Recent experience in Britain suggests that there is virtually no scope for local authorities to pursue policies that are at variance with central government's wishes. When the county of South Yorkshire persisted with a cheap fares policy for public transport, central government withdrew certain transport grants due to the county. Even in the United States, where one might expect a lesser degree of control by the federal government, progress on new rapid transit systems is effectively controlled by the Urban Mass Transit Administration, a federal agency with powers to fund transit proposals made at a regional or local level.

TECHNOLOGICAL CHOICE

Once a decision has been made to invest in rapid transit, a choice has to be made among the available technologies. Should the money be spend on conventional railways or new systems such as monorails and busways? While this choice could be regarded simply as a technological evaluation, it is often linked with the basic decision whether to build rapid transport or not.

In both Britain and France, the compatibility of established and new systems has been an important factor in the choice of technologies, particularly where any new system has to be grafted on to a long established network of railway or metro lines. These difficulties obviously relate to engineering compatibility, but a study of the convenience of passengers is also required. Thus, when the extension of the Paris metro line to Creteil was proposed, the use of the Safege monorail for the extension would have entailed an inconvenient change of vehicle at the metro terminal at Charenton. Likewise, the advantages of making any new system compatible with the existing suburban railway network emerged at an early stage of the Manchester Rapid Transit Study.[7]

The outcome of investigations into available technologies has undoubtedly been influenced by the nature of the organisation making the evaluation. In the case of the Paris metro extension, the technical evaluation was undertaken by the Régie Autonome des Transports Parisiennes (RATP), and in the Manchester Transit Study by North American railway consultants. It is not perhaps surprising that in both cases the conclusions favoured the system most familiar to the body making the study. More serious, as Hamer has pointed out in relation to North American studies, is the danger that the objectivity of consultants may be influenced by the fact that they could profit from a decision to go ahead with rapid transit.[2] However, these are difficulties which are hard to avoid, for often the people best qualified to give advice on a subject are those who have some prior experience of that subject!

Differences between the environmental impacts of the available technologies do not seem to have had any major effect on the choice of technologies evaluated for the Paris metro extension, which ran largely through undeveloped land. On the other hand, in Britain one of the advantages initially claimed for the Safege monorail was that it could be fitted into an established city with less environmental impact than a conventional railway. The Manchester Rapid Transit Study found this not to be so.[7]

LOCATIONAL AND BEHAVIOURAL IMPACTS

There may be two kinds of effects on land use: locational choices made by developers who expect to benefit; and actual changes in the use made of buildings, following changes in travel behaviour. These changes in travel behaviour may or may not be substantial. In the case of offices built adjacent to the extension of the Paris metro line to Creteil, the use of the new metro by office employees proved to be slight. On the other hand, the management of the hypermarket was surprised to find that some 10 per cent of its customers arrived by metro. An even smaller impact had been anticipated. More important has been the use of the metro to get to the new hospital and medical school. The extent of the impact may depend on the pace of urbanisation. The massive development of land around the Toronto rapid transit lines reflects in part the rapid urbanisation of the region.

No comparable situation has arisen in the UK, simply because what new railway construction has taken place has not been on undeveloped land on the fringes of the urban area. The investment in Glasgow, Liverpool and Newcastle has been concentrated on the provision of cross-centre links, joining previously unconnected routes. In these cases, the bulk of the route mileage is on long established suburban rail routes, with a short length of new and modernised tunnels beneath the city centre. However, where plans for upgraded suburban railway lines have been prepared, planners have anticipated that land-use changes would follow. Thus, the Birmingham structure plan originally proposed that redevelopment of housing around suburban railway stations should be permitted to take place at a higher density than in other locations.[8] It was argued that the best use would be made of investment in improved suburban railways and roads would be less congested. This proposal led to public protest, however. The residents of attractive, low-density outer suburbs did not like the prospect of a change in character of their areas, and the Secretary of State rejected the proposal when the structure plan was submitted to him for approval.

Perhaps a more serious criticisim is that, even if the measures were implemented for all redevelopment sites available within walking distance of improved suburban railways in the conurbation, the overall effect on traffic volumes perceived by road users would be negligible. The further radial extensions of metro lines extend from the centre, the greater their potential usage. Within an expanding suburban region, however, the proportion of the population within easy access of a rapid transit station will diminish. This provides one obvious explanation why in other suburban areas, roads, such as the tangential route RN 186 to the south of Paris, rather than railways, can be seen as foci for urban development, even in those city regions where strenuous efforts are being made to improve the quality of rail-based public transport.

ECONOMIC AND SOCIAL CRITERIA FOR SELECTION AND EVALUATION

In practice, many decisions to go ahead or not with proposals for new rapid transit have been influenced by political factors, narrow economic assessments and indeed accidents of history. Is a more broadly based assessment method possible? In a recent paper, Sagner and Barringer comment that in the search for criteria for the selection of rapid rail transit capital fund recipients, planners have sub-optimised by failing to consider critically the wider ramifications of such investment decisions.[9] Yet the same paper goes on to concentrate wholly on the effects of rapid transit on the concentration of employment. Admittedly, this appears to be one of the most tangible possible effects of rapid transit construction.

The example of Toronto, where 50 per cent of high-rise construction was located on the new subway line, is widely quoted. In Los Angeles it was estimated by the Stanford Research Institute that if rapid transit were made available the monthly welfare rolls would be reduced by 4,200.[10] An example quoted was the garment manufacturing industry, in which it was estimated there would be a 60 per cent increase in availability of labour. Even if we assume this to be an accurate forecast, such employment consequences of new transit are open to varying interpretations. Improvements in public transport might encourage more housewives to go out to work, either full or part time. This could be regarded as an economic benefit, representing an increase in activity rates, or it could be seen by workers already employed in the industry - a low-wage industry - as a means whereby employers can hold down wage rates.

Thus, in attempting to anticipate the consequences of a new rapid transit line, it is necessary to ask whether new jobs will be created or merely redistributed. Moreover, it is questionable whether expected redistribution is desirable. There can be little doubt that the effect of the new metro in Amsterdam in stimulating office development around city centre stations has had a very destructive impact on the character and human scale of the old city.

The effect of rapid transit in making central areas more accessible and hence more attractive to developers is one of the most widely anticipated consequences. Indeed, the main stimulus to growth in central area economic activity and the retention, if not increase, of service and retailing employment may occur in the period after a decision to build a new rapid transit line and before its actual opening. For example, office space in central San Francisco increased by 78 per cent between 1962 and 1969, with all the new buildings within five minutes of a proposed BART station.[11] Even so, doubts remain about whether there was a causal relationship between the BART proposal and the property boom. Los Angeles, Seattle and Portland experienced similar office construction booms during the 1960s without the benefits claimed for rapid transit.

While it is possible to dispute the evidence of the effects on central business districts of new rapid transit construction, it remains true that one of the most important non-quantitative arguments in favour of rapid transport is this. The increased use of motor traffic and the building of regional motorway networks have made long established city centres relatively less accessible. One way to redress this is by improved public transport access to the centre, i.e. a planning decision is necessary - with the institutions making the decision varying from country to country - to preserve the vigour of established city centres. This perhaps represents the opposite extreme to a narrow economic cost-benefit evauulation of a transport investment proposal. Even to consider using such broad planning goals as a basis for decision making means that many new questions will be asked. How important is service and retailing employment to the regional economy? For whose benefit is the city centre to be made more accessible? Are only journeys to work important? How should environmental gains be measured?

The distributional effects of rail rapid transit construction are still relatively little researched. It is possible to make the general statement that rapid transit is most suitable for the longer commuter journeys from outer suburbs to city centre, and that the outer suburbs are typically inhabited by the higher income groups. The 1977 transport white paper (HMSO) recognised that a subsidy to support improved commuting facilities could turn out to be a subsidy to the wealthy.[12] Thus, the new central area loop line in Liverpool shortened the journeys for the wealthier commuters from the Wirral. Yet generalisations are difficult. The Tyneside metro, now under construction, links the city centre of Newcastle with a number of roughly linear settlements on each side of the river Tyne. As the population of the region has not grown on any monocentric radial pattern, only about eight or nine of the forty-four stations on the system could be classified as within upper income areas.

In looking at who benefits from new rapid transit, it is perhaps simplistic to look only at the income of the potential users. If one directs attention to the "transportation poor", it is evident that it is not possible to locate a transit line specifically to benefit such groups. Old people, whether deprived of the mobility of car ownership by reason of infirmity or low income, are not located within any narrowly defined geographical areas. Similarly, children below the age of seventeen, who have travel needs but no access to cars, are similarly scattered. While it is possible to identify areas with low-income inhabitants it must be borne in mind that a new rapid transit line should have a life of sixty years or more; so to use rapid transport as a means of positive discrimination in favour of certain groups would be a very indirect and inflexible public policy.

The main transport need of old people is for access to local facilities, shops in particular. Clearly, as rail rapid transit is best suited to the longer urban journeys, it cannot be seen as a particular transport benefit for this section of the "transportation poor". Similarly, as far as children are concerned, improvements in short local journey conditions would be of most benefit. For certain recreation trips, e.g. to large-scale spectator sports, rail-based public transport could be an appropriate way of effecting travel benefits, even if they would not be directed to any specific groups in society. Of the total users of the BART, only 6 per cent are under 18 or over 65, while these two groups constitute 23 per cent of the area's population.[2]

In considering who benefits from public transport improvements, it is important to distinguish between rapid transit and bus services, for there is no doubt about the reliance placed on bus services by the elderly and lower income groups. Unfortunately, the history of bus operations as self-financing and even profitable organisations has made bus operators slow to adopt broadly based evaluation methods to assess possible improvements. Indirect benefits from changes tend to be dismissed as "fairy gold", for the benefits do not appear on balance sheets, whereas extra costs in operating a more frequent service are readily apparent. Operators are dubious, perhaps rightly, about whether their passengers will perceive small time savings as a benefit. Yet the fact remains that urban road building schemes are conventionally justified on the basis of small time savings for large numbers of drivers.

SOME LESSONS FOR THE FUTURE

Perhaps one of the main lessons to be learned from the recent history of rail rapid transit investment decisions is that the problems of assessing indirect social efforts of capital-intensive schemes are not easily tractable, and that more effort should be directed towards ensuring a consistent basis for evaluating proposals for major road schemes, rail-based schemes and lower cost measures for improving bus public transport, where the main costs may be operational rather than capital, and the main benefits are either quantifiable or can be seen to be received by an identifiable group in society.

Clearly, in conditions of rapid urbanisation and population growth, the building of rapid transit lines had an enormous influence on the distribution of land values and on the form, use and intensity of land buildings, within and between cities. Much recent investment in rapid transit has taken place within urban areas, and there is little doubt that similar effects will be perceived over time, depending on the scale of the facility. Where transit acts as a catalyst for change in the social and economic well-being of areas within a city, the ramifications of second order effects are manifold. An increase in property values may be seen by some as a benefit, by others as perhaps destructive of the existing social cohesion within the locality. Once the attempt is made to order effects into any project evaluation, the task is made enormously difficult. One might observe that the construction of the Victoria line underground in London has been accompanied by social changes and an increase in property values in certain areas north-east of central London, which are traversed by the new route. No such evident social changes have occurred around the southern end of the line at Stockwell and Brixton, however. Some economic and social changes may become more predictable once appropriate evaluation procedures have been developed. Nevertheless, it remains difficult to predict who will benefit from any forecast changes, such as urban renewal occurring around an inner city station, once a decision has been made to develop rapid transit. It also remains difficult to establish causality.

A more fruitful approach might be based on the evaluation of welfare changes at the micro level, i.e. within particular localities, where particular transportation-related problems can be identified.

Such an approach needs more careful investigation. It implies the establishment of welfare and equity objectives for the delivery of goods and services, and quality control mechanisms analogous to other public goods. The difficulties arise in assessing the social choice criteria to be adopted in any particular case. As we move away from long-term designed high technology transit solutions to more flexible, responsive and diverse bundles of policies, likely to respond better to problems which various groups encounter in gaining access to the myriad activities of the city, we will need to develop new decision frameworks and measures for learning whether such policies are achieving their objectives.

The local economy in many parts of UK city regions has suffered from long-run trends and structural changes exacerbated by regional policy, implementation of which has included the developments arising from the location of major transport infrastructure and it is evident that this has led to primary and secondary impacts on land use.

It is not unreasonable to suggest that if other activities locate at points of maximum accessibility, economies of agglomeration may induce growth poles at planned locations on the urban periphery. Such growth may be of critical importance for a local economy and would conflict with containment and green belt policies operating on the periphery of UK urban regions. But such growth may provide much needed employment and less social problems in the inner older areas. This final and controversial observation suggests that we have hardly begun to explore the planning balance sheet that must be drawn up and the trade-offs that have to be made before we can state that the criteria for planned investment in rapid transit or indeed any other transport mode include such economic and social factors.

REFERENCES

1   White, P.R., Planning for public transport (1976) Hutchinson.
2   Hamer, A.H., The selling of rail rapid transit (1976), Lexington.
3   Dupuy, G., Une technique de planification au service de l'automobile: les models de trafic urbaine (1975), L'Institute d'Urbanisme de Paris.
4   Thomson, J.M., "Half way to a Motorised society", Problems of an urban society (1973) Vol 3, ed. J.B.Cullingworth, George Allen and Unwin.
5   Thoenig, J.C., and Despicht, N., "Transport policy, Chapter 6", Planning politics and public policy: the British, French and Italian experience (1975), ed. J. Hayward and M. Watson, Cambridge University Press.
6   Starkie, D.N.M., Transportation planning and public policy: progress in planning (1973) Vol 1, Part 4, Pergamon.
7   De Leuw Cather and Partners, Manchester Rapid Transit Study (1967), Manchester Corporation.

[8]   City of Birmingham structure plan: written statement (1973) Birmingham Corporation
[9]   Sagner, J.S., and Barringer, R.C., "Toward criteria in the Development of urban
        transportation systems" (1978), Transportation, Vol 7, No 1.
[10]   Stanford Research Institute, Benefit/cost analysis of the five-corridor rapid Transit
        system for Los Angeles (1968).
[11]   Sheldon, N.W. and Brandwein, R., The economic and social impact of investments in
        public transit (1973), Lexington.
[12]   HMSO White Paper. Transport Policy (1977), Cmnd 6836.

An entropy-maximising distribution model and its application to a land-use model

7

Katsunao Kondo and Tsuna Sasaki   Japan

SYNOPSIS

The purpose of this paper is to focus on the use of entropy models in the analysis of travel and spatial interaction. First, the paper reviews entropy models in trip distribution, particularly from the point of view of solution structure. The most likely assignment of trips to given origins and destinations appears to be a purely statistical problem. It is solvable as a statistical problem only when we introduce one additional piece of information - say, a priori probability between zones, or total travel expenditure in a study area. The former model has been proposed by Sasaki in 1968 and the latter by Wilson in 1967. Here we introduce an entropy-maximising model which would offer a better understanding of the structures of two conventional entropy models. We also touch briefly on the basic ideas underlying them. A careful analysis of balancing factors in this new method shows that the entropy model of the current pattern type is in every way equal to the Detroit model, which is one of the growth factor techniques.

The remainder of this paper is concerned with model-based methods of estimating and forecasting the distribution of urban activities in a city, particularly the numbers of households. The equation which locates the population in each zone on the basis of employment is derived from the entropy-maximising trip distribution model of the gravity type. This equation indicates the most probable pattern of household locations on the assumption that trip distribution forms a gravity pattern.

INTRODUCTION

The entropy-maximising technique is now popularly used in the analysis of spatial interaction in transport studies, economics and regional studies. Though the gravity models plays an important role as an empirical model for spatial interaction, much has been written on the relation between the gravity model, utility model and entropy model. The main concern of those interested in theoretical work has been to derive the gravity pattern theoretically. Once this is done, they do not give more thought to the assumed behavioural theory or the derived solution of the gravity pattern. As a matter of fact, the derived theoretical pattern of the gravity model has some important aspects. First, this paper describes the entropy model of current type which will offer a better understanding of the solution structure of conventional entropy models of the gravity type. Careful analysis of the balancing factors in this new method shows that the entropy model of current pattern type is in every way equal to the Detroit model, which is one of the growth factor techniques.[1] Snickars and Weibull also investigated this problem, but not sufficiently.[2] Murchland has discussed the existence and uniqueness of the solutions in double-balancing problems.[3] [4] We also mention such a problem in this paper.

The remainder of this paper is concerned with model-based methods of estimating and forecasting the distribution of households in a metropolitan area. The so-called Lowry model,[5] an activity-allocation model originally developed for US metropolitan areas, is too simple to describe actual urban activity locating patterns. Thus, we try to reform its key equation, say, the accessibility equation, on the assumption that the OD pattern of journey-to-work trip forms a gravity pattern. Suburbanisation and multi-centred urbanisation can be reflected in an attracting coefficient which appears in the new equation for accessibility.

STATISTICAL INTERPRETATION OF GROWTH FACTOR TECHNIQUE

In growth factor techniques (present pattern methods) the number of trips between zones i and j is estimated only by using a given OD table for the base year and the growth factors for zones i and j which reflect the growth in trip productions and trip attractions expected between the base and horizon years. Friction factors between zones are, of course, not taken into account (the friction factor in the horizon year is assumed to be the same as that of the base year). From the statistical point of view, we may give the statistical interpretation to the growth factor technique by introducing the condition that the OD pattern of the horizon year should be that most similar to that of the base year. Needless to say, the concept of "most similar" includes various kinds of optimisation problems. The familiar techniques of this kind are as follows: minimisation of the residual sum of squares, minimisation of qui squares, and maximisation of joint probability of occurrence. The former two models can be studied in statistics textbooks. Here we consider only the latter model.

Before describing the model we must assume one prior piece of information. That is that "a priori probability of the (i, j)th cell in the horizon year OD table must be equal to the unit OD table of the base year".

This assumption can be written as follows:

$$p'_{ij} = X'_{ij}/T' \qquad \text{for all } (i,j) \tag{2.1}$$

where p'_{ij} is the a priori probability in the horizon year OD matrix, X'_{ij} is the number of trips between zones i and j in the base year, and $T' = \Sigma'_{ij} X'_{ij}$. X'_{ij}/T' can be called the unit OD table.

Now we consider a joint probability that traffic flows of X_{ij}'s are distributed to the (i, j)th cell of OD matrix. By polynomial theorem, the joint probability is written as follows:

$$P = \frac{T!}{\Pi X_{ij}!} \Pi (p'_{ij})^{X_{ij}} \tag{2.2}$$

where X_{ij} represents the number of trips between zones in the horizon year, and $T = \Sigma_{ij} X_{ij}$. Then our problem is expressed as follows: maximise P, subject to $\Sigma_j \; X_{ij} = U_i \qquad (i=1,2,\ldots,N),$ (2.3)

$\Sigma_i \; X_{ij} = V_j \qquad (j=1,2,\ldots,N),$ (2.4)

where U_i represents trip productions in zone i and V_j trip attractions to zone j. Both U_i and V_j are given. Maximising P is equivalent to maximising ln P, so our problem can be rewritten as follows by using Lagrangian multipliers, after omitting constant terms and also using Stirling's approximation (ln $X = X$ ln $X - X$):

maximise L

$$L = \Sigma_{ij} \; X_{ij} \ln p'_{ij} - \Sigma_{ij} \; (X_{ij} \ln X_{ij} - X_{ij})$$
$$+ \; \Sigma_i \alpha_i \; (\; \Sigma_j X_{ij} - U_i) + \Sigma_j \beta_j (\; \Sigma_i X_{ij} - V_j)$$

where the α_i's and β_j's are Lagrangian multipliers. The X_{ij}'s, which maximise L and therefore constitute the most probable distribution of trips, are the solution of

$$\partial L \; / \; \partial \; X_{ij} = 0$$ (2.5)

and the constraint equations (2.3) and (2.4). Equation (2.5) gives us the solution of

$$X_{ij} = \exp \; (\alpha_i + \beta_j) P'_{ij}.$$ (2.6)

To obtain the final result in more familiar form, write

$$\lambda_i = \exp \; (\alpha_i) \text{ and } \mu_j = \exp \; (\beta_j)$$

and then we have

$$X_{ij} = \lambda_i \mu_j p'_{ij}$$ (2.7)

where λ_i and μ_j are called balancing factors. Substituting X'_{ij}/T' in spite of p'_{ij}, and writing

$$\overset{*}{\lambda}_i = \lambda_i / T' \text{ and } \overset{*}{\mu}_j = \mu_j,$$

Equation (2.7) and constraint equations can be written as follows:

$$X_{ij} = \overset{*}{\lambda}_i \overset{*}{\mu}_j \; X'_{ij},$$ (2.8)

$$\overset{*}{\lambda}_i = U_i \; / \; \Sigma_j \; \overset{*}{\mu}_j X'_{ij},$$ (2.9)

$$\overset{*}{\mu}_j = V_j \; / \; \Sigma_i \; \overset{*}{\lambda}_i X'_{ij},$$ (2.10)

Thus, the most probable distribution of trips is expressed by using balancing factors $\overset{*}{\lambda}_i$ and $\overset{*}{\mu}_j$. This is called double-balancing problem by Murchland, who has mentioned the existence and the uniqueness of balancing factors.[3][4] It should be noted, however, that we do not need the unique solutions for $\overset{*}{\lambda}_i$ and $\overset{*}{\mu}_j$ in order to estimate X_{ij}. For our purpose, the value of $\lambda_i^* \mu_j^*$ ($= f_{ij}$), growth factor of the (i, j)th traffic, must be unique. As a matter of fact, the uniqueness of f_{ij} can be shown as follows.

We have N^2 variables (f_{ij}'s), and also $2N-1$ conditions [Equation (2.9) and Equation (2.10)], Equations (2.9) and (2.10) are not independent of each another, because there is the relation that $\Sigma U_i = \Sigma V_j = T$. However, we can prepare additional conditions for f_{ij}'s of N^2-2N+1 through the relation of $f_{ij} = \lambda_i^* \mu_j^*$.

Write $F_{11}:f_{12}: \cdots\cdots\cdots :f_{1N} = f_{21}:f_{22}: \cdots\cdots\cdots :f_{2N}$

$= f_{31}:f_{32}: \cdots\cdots\cdots :f_{3N}$

$\cdots\cdots\cdots\cdots\cdots\cdots$ $N-1$ (2.11)

$= f_{N1}:f_{N2}: \cdots\cdots\cdots :F_{NN}$

$(= \mu_1^*: \mu_2^*: \cdots\cdots\cdots : \mu_N^*)$.

These $N-1$ proportional equations can produce $(N-1)^2$ micro-proportional equations:

$$f_{11}:f_{12}=f_{21}:f_{22} \qquad f_{11}:f_{13}=f_{21}:f_{23} \qquad f_{11}:f_{1N}=f_{21}:f_{2N}$$
$$= f_{31}:f_{32} \qquad\qquad =f31:f_{33} \qquad\qquad =f_{31}:f_{3N}$$
$$\cdots\cdots \qquad\qquad \cdots\cdots \qquad\qquad \cdots\cdots \qquad (2.12)$$
$$=f_{N1}:f_{N2'} \qquad\qquad =f_{N1}:f_{N3'} \qquad\qquad =f_{N1}:f_{NN'}$$

Thus, the number of the additional conditions is $(N-1)^2$, which is equal to N^2-2N+1.

In addition, we can prove that the solutions to the double-balancing problem mentioned above are completely the same as those of the Detroit model. In other words, the iteration procedure for solving Equations (2.9) and (2.10) is equivalent to that of the Detroit model. The Detroit model can be expressed as follows:

$$X_{ij}^{(1)} = X_{ij}' F_i G_j / H \qquad\qquad (2.13)$$

where X_{ij}' is the base year's OD matrix, F_i is the growth factor for trips produced in zone i, G_j is the growth factor for trips attracted to zone j, H is the growth factor for trips generated in a study area, and $x_{ij}^{(1)}$ is the first approximation of the horizon year's OD matrix. In the Detroit model, iteration must be carried out until the values of the following recalculated growth factors approach one:

$$F_i^{(1)} = U_i / \Sigma_j X_{ij}^{(1)} \,, \qquad\qquad (2.14)$$

$$G_j^{(1)} = V_j / \Sigma_i X_{ij}^{(1)} \,. \qquad\qquad (2.15)$$

Essentially, the growth factor H is not important in Equation (2.13), because there is no suffix i or j. Thus, we omit H in the following discussion. From Equations (2.13), (2.14) and (2.15), we have

$$F_i F_i^{(1)} = U_i / \Sigma_j X_{ij}' G_j \,, \qquad\qquad (2.16)a$$

$$G_j G_j^{(1)} = V_j / \Sigma_i X_{ij}' F_i \,. \qquad\qquad (2.16)b$$

If iteration ends at the n-th step, this iteration process can be written as

$$F_i F_i^{(1)} F_i^{(2)} \ldots F_i^{(n)} = U_i / \Sigma_i X'_{ij} (G_j G_j^{(1)} G_j^{(2)} \ldots G_j^{(n)}),$$ (2.17)

$$G_j G_j^{(1)} G_j^{(2)} \ldots G_j^{(n)} = V_j / \Sigma_i X'_{ij} (F_i F_i^{(1)} F_i^{(2)} \ldots F_i^{(n)}).$$ (2.18)

Compare these Equations (2.17) and (2.18) with Equations (2.9) and (2.10). We can expect that both iterations will reach the same solutions. According to our calculation, as a matter of fact, both iterations reached the same solutions, as mentioned in the Appendix. That is to say

$$\lambda_i^* = F_i F_i^{(1)} F_i^{(2)} \ldots \ldots \ldots F_i^{(n)},$$

$$\mu_j^* = G_j G_j^{(1)} G_j^{(2)} \ldots \ldots \ldots G_j^{(n)}.$$

The only difference between the entropy model and the Detroit model lies in the technique for the convergence criterion. Therefore, it is concluded that the entropy-maximising model of the present pattern type given the theoretical basis for the Detroit model and also for the double-balancing problem. This point has not been sufficiently examined. There is no doubt that this study will assist in acquiring a better understanding of the trip distribution model by both transport engineers and planners. As for equations (2.9) and (2.10) we do not necessarily have to solve both λ_i^* and μ_j^*, because either λ_i^* or μ_j^* can be eliminated as follows:

$$X_{ij} = U_i \frac{\mu_j^* X'_{ij}}{\Sigma_k \mu_k^* X'_{ik}}$$ (2.19)

sub. to $\Sigma_i X_{ij} = V_j$ (j=1, 2, ... ,N). (2.20)

Such a solution is more familiar to us. Our first form of solution expressed in Equation (2.8) was doubly constrained, but we have now obtained a singly constrained solution as above. Thus, the most probable distribution of trips is expressed by using balancing factors μ_j^*'s, and Equation (2.19) is the last form of the solution. Double-balancing problems, therefore, can be changed to single-balancing problems. Murchland,[3][4] mentioned that the double-balancing problem always has a solution if the matrix X'_{ij} has no zero entries. As mentioned above, however, a single-balancing problem, as in Equation (2.19), always has a solution if some X'_{ij}'s have a zero value. This is obvious from the Equation (2.20), which can be rewritten as

$$\mu_j^* = V_j / \Sigma_i (U_i X'_{ij} / \Sigma_k \mu_k^* X'_{ij}).$$ (2.21)

SOME ASPECTS OF THE CONVENTIONAL ENTROPY MODEL OF THE GRAVITY TYPE

Conventional entropy-maximising distribution models of the gravity type are basically classified into three types: one using a priori probability of the gravity type, a second using the third-constraint equation on total travel expenditure, and, finally, the integrated model which combines these two models. The first was proposed by Sasaki,[6] the second by Wilson,[7] and the third by Snickars and Weibull.[2] We shall now briefly touch on the basic concepts of the first two models. The third one, the integrated model, seems to be one of the possible extensions of the two conventional models. Since the statistical theory of the entropy-maximising model is discussed in depth in the paper by Snickars and Weibull, we focus on the structure of solutions by the two conventional models.

The formulation of Sasaki's model is the same as that of the entropy model of the present pattern type discussed earlier in this paper. The form of a priori probability in his model was originally assumed to be

$$p'_{ij} = \kappa\, u_i v_j\, t_{ij}^{-\gamma} \tag{3.1}$$

where $u_i = U_i/T$ (producing power), $v_j = V_j/T$ (attracting power), t_{ij} is travel time between zones i and j, γ is a constant which should be determined by observed data, and κ is a constant which ensures $\Sigma_{ij}\, p'_{ij} = 1$. New findings on his model are introduced below. The form of the solution of his model is, of course, the same as that of Equation (2.7), i.e.

$$X_{ij} = \lambda_i \mu_j\, p'_{ij}. \tag{3.2}$$

If we transfer p'_{ij} in Equation (3.1) to Equation (3.2) and write

$$\overset{*}{\lambda}_i = \kappa\, \lambda_i u_i \qquad \text{and} \qquad \overset{*}{\mu}_j = \mu_j v_j \quad,$$

we get

$$X_{ij} = \overset{*}{\lambda}_i\, \overset{*}{\mu}_j\, t_{ij}^{-\gamma} \tag{3.3}$$

where $\overset{*}{\lambda}_i$ and $\overset{*}{\mu}_j$ are redefined balancing factors. Of the terms which form a priori probability in Equation (3.1) only the term $t_{ij}^{-\gamma}$ appears in the solution of Equation (3.3). Other terms related to zones i and j can be included in the new balancing factors. This means that the most important factor in the equation of a priori probability is a travel time function between zones. As a matter of fact, we can write the generalised form of a priori probability as a gravity form by

$$p'_{ij} = \kappa\, u_i^{\,\alpha}\, v_j^{\,\beta}\, f(t_{ij}) \quad, \tag{3.4}$$

but the solution takes the form

$$X_{ij} = \overset{*}{\lambda}_i\, \overset{*}{\mu}_j\, f(t_{ij}) \quad. \tag{3.5}$$

The terms related to zones i and j can be included in the balancing factors. Moreover, using the constraint equations (2.3) and (2.4), the solution of Equation (3.5) can be written in a more familiar form as

$$X_{ij} = U_i\, \frac{\overset{*}{\mu}_j\, f(t_{ij})}{\Sigma_k\, \overset{*}{\mu}_k f(t_{ik})} \qquad \text{sub. to } \Sigma_i\, X_{ij} = V_j \tag{3.6a}$$

or

$$X_{ij} = V_j \frac{\overset{*}{\lambda_i} f(t_{ij})}{\Sigma_k \overset{*}{\lambda_k} f(t_{ij})} \qquad \text{sub. to } \Sigma_j X_{ij} = U_i. \qquad (3.6)b$$

The solution given by Equation (3.6)a is equal to that of the often used gravity model, in which the $\overset{*}{\mu}$'s are called the trip-attraction magnitudes transferred to Equation (3.6)a for a successive iteration of the model. Therefore, it is concluded that the entropy- maximising model using a priory probability of the gravity type can offer a theoretical basis for the conventional gravity model iteration procedure.

A trip distribution by the gravity model must satisfy the constraint Equations (2.3) and (2.4) but Wilson,[6] has introduced a third constraint, which is

$$\Sigma_{ij} c_{ij} X_{ij} = C$$

where c_{ij} is the generalised cost of travelling from zone i to j and C is the total amount spent on travel in the region at a given time. Wilson has shown that the most probable X_{ij} matrix to satisfy the three constraints mentioned above is given by the following form of the gravity model:

$$X_{ij} = \alpha_i U_i \beta_j V_j \exp(-\Upsilon c_{ij}) \qquad (3.8)$$

where α_i and β_j are balancing factors. If we write $\overset{*}{\alpha_i} = \alpha_i U_i$ and $\overset{*}{\beta_j} = \beta_j V_j$, then we have

$$X_{ij} = \overset{*}{\alpha_i} \overset{*}{\beta_j} \exp(-\Upsilon c_{ij})$$

and finally

$$X_{ij} = U_i \frac{\overset{*}{\beta_j} \exp(-\Upsilon c_{ij})}{\Sigma_k \overset{*}{\beta_k} \exp(-\Upsilon c_{kj})} \qquad \text{sub. to } \Sigma_j X_{ij} = V_j \qquad (3.9)$$
$$\text{and Equation (3.7)}$$

This form of the solution is equal to Equation (3.6)a. Therefore, his model is thought to be one offering a theoretical basis for the gravity model. His model must be discussed from the following points of view, however: the meaning of the third constraint equation, and the feasibility of the empirical test. As for the first point, consider the entropy-maximising model without the third constraint, that is

$$\text{max.} \quad \frac{T!}{\Pi X_{ij}!} \qquad \text{sub. to (2.3) and (2.4)}$$

The solution of this problem is quite simple:

$$X_{ij} = U_i V_j / T \qquad (T = \Sigma U_i = \Sigma V_j). \qquad (3.10)$$

Therefore the third constraint of Equation (3.7) adds a friction factor between zones to the solution of Equation (3.10). As a matter of fact, Equation (3.7) only means that the total expenditure on travel in the region must be equal to the total trip cost of all origin-destination pairs. Once Equation (3.7) is taken into account as a condition, however, it acts as a constraint on X_{ij}. There is no theoretical or empirical basis for Equation (3.7); it is only an assumption. Therefore, the solution of Equation (3.8) or (3.9) is a purely theoretical one. This critique also applies to the second point (b). In Wilson's model, the parameters $\overset{*}{\beta}_j$'s and Υ, which form the solution of Equation (3.9), should be determined so that the constraint equations are satisfied. There is no problem as to $\overset{*}{\beta}_j$'s, because there are balancing factors. There remains a question about the parameter Υ, however. Parameter Υ in the conventional concept is a factor which represents the distribution of trip length. Therefore, according to the conventional concept, parameter Υ must reflect the real distriburion of trip length and must be determined by observations. In Wilson's model, however, parameter Υ is determined by the constraints. It should be noted that this method for determining Υ can be applied only in the base year, because we do not know future expenditure on travel. Consequently, the Wilson model is a tool for explaining the present OD pattern theoretically. On the other hand, the Sasaki model can be tested by observed trip data. Parameter Υ on t_{ij} can be determined positively (by the least-squares method for Equation (3.1)). We can also test various kinds of a priori probability, for instance Equation (3.4). Therefore the Sasaki model has greater elasticity, compared with the Wilson model. The planner can choose the most suitable form of a priori probability by statistical test. In the Wilson model, there is no room for the planner to select the most reasonable OD pattern from the point of view of suitability.

We have now reviewed two conventional entropy-maximising distribution models. In the next section we will introduce model-based household location methods on the assumption that journey-to-work trips take a gravity pattern. Lowry's household location model will be developed by using an entropy model of the a priori probability type.

THE RESIDENTIAL LOCATION MODEL

The purpose of this section is to improve the residential location equation in the so-called Lowrey model[5]. Other aspects and possible extensions of the Lowry model have been well discussed by Wilson.[8] As is well known, the Lowry model has a particularly well-defined causal structure, as exhibited in Figure 1. The allocation of basic employment is assumed to be given. The households of these workers are then assumed to be located around these work places. This population then generates a demand for retail activity, which is assumed to be located in relation to the population. These additional jobs then generate more population, and the new population an additional demand for retail services, and so on. This explains the iterative loop in Figure 1.

The key equation in the Lowry model is a potention equation for household location, which is expressed by

$$N_i = g \, \Sigma_j \, E_j / t_{ij} \qquad\qquad (4.1)$$

sub. to $\Sigma_i \, N_i = N$

where N_i denotes the population in zone i, E_j denotes the employment in zone j, t_{ij} is a friction factor between zones i and j, N is the total population in a metropolitan area, and g is a constant which ensures $\Sigma_i \, N_i = N$.

Equation (4.1) seems to be extremely simple and does not reflect the OD pattern of journey-to-work trips. This model was originally developed in order to apply it to US metropolitan areas. Therefore, it cannot explain recent patterns of residential location.

Lowry assumed the one-cored urban structure and the closed urban system. These assumptions are not valid today. Big city and metropolitan areas now have sub-city centres and multi-cores. Rapid suburbanisation and enlargement of behavioural space make it impossible to persist in the assumption of closed systems. We must therefore improve the original model to reflect these changes. In order to do this, it is necessary to introduce the attractiveness of living in each zone and the OD pattern of journeys to work. We develop the residential location model with the entropy model, using the a priori probability below.

In terms of the Lowry model, say Equation (4.1), E_j's, t_{ij}'s and N are now given. Then we introduce additional information on the attractiveness of living in each zone, which is expressed by w_i. This attractivenss may include the ability to live in a zone, the number of houses, etc. Then we assume the prior probability for journey-to-work trips as a gravity form

$$p'_{ij} = \kappa \; w_i \; E_j \; t_{ij}^{-\Upsilon} . \tag{4.2}$$

Then, the most probable distribution of trips may be obtained by solving the following maximising problem:

$$\text{max.} \quad \frac{E!}{\Pi \; X_{ij}!} \; \Pi p'_{ij}{}^{X_{ij}} \qquad\qquad \text{sub. to } \Sigma_i \; X_{ij} = E_j$$

where $E = \Sigma \; E_j$ and X_{ij} the journey-to-work trips between zones i and j. At the same time it indicates the number of workers in zone j living in zone i.

The solution of the problem can be written as

$$X_{ij} = E_j \; \frac{w_i \; t_{ij}^{-\Upsilon}}{\Sigma_k \; w_k \; t_{kj}^{-\Upsilon}} . \tag{4.3}$$

The technique for solving the problem is the same as that used earlier. It should be noted, however, that there is no constraint equation on the trip production in each zone. Using Equation (4.3) we can calculate the number of workers living in zone i, that is

$$L_i = \Sigma_j \; X_{ij}$$

$$= \Sigma_j \quad E_j \quad \frac{w_i \; t_{ij}^{-\Upsilon}}{\Sigma_k \; w_k \; t_{kj}^{-\Upsilon}} \tag{4.4}$$

Then we obtain the population in zone i as follows:

$$N_i = f \; L_i \tag{4.5}$$

where f is the inverse of the activity rate, and so represents the ratio of population to employed residents in the study area.

Equation (4.4) is basically important. Such a form of the solution cannot be obtained by the model of the Wilson type. Only the entropy model of the Sasaki type can produce such a familiar and acceptable form of solution.

When we forecast the L_i's, the most important task is to specify the form of the attractiveness w_i. As is evident from Equation (4.4), the real value of w_i is not so significant, but the relative value is of interest. If we can specify and forecast the w_i's with some degree of accuracy, the population in each zone can be estimated by Equations (4.4) and (4.5).

The remainder of this section studies the residential locations at three points in time in the Osaka metropolitan area, the second largest in Japan. Osaka city is located in the centre of this metropolitan area and its population has been decreasing for the past ten years, while the surrounding areas have been rapidly and continuously suburbanised. Equation (4.4) is applied to such a dynamic region below.

Equation (4.4) may be changed as follows in order to calculate the values of w_i's:

$$w_i = L_i \left/ \; \Sigma_j \left(\frac{t_{ij}^{-\Upsilon}}{\Sigma_k \, w_k \, t_{kj}^{-\Upsilon}} \right) \right. \tag{4.6}$$

Using the given values of L_i's, E_j's, t_{ij}'s and Υ at a particular time, we can obtain the values of w_i's. As mentioned above, the value of w_i is not of interest to us. Then we prepare the attracting coefficient as follows:

$$\overset{*}{w}_i = w_i \, \frac{22}{\Sigma \, w_i}$$

where 22 is the number of zones in Osaka. This new index for the attractiveness of living in each zone means that if the attractiveness of zone i is equal to one-22nd of the total amount $\Sigma \, w_i$, the attracting coefficient $\overset{*}{w}_i$ takes the value of one. Such an index makes it possible to compare the attractiveness of zones at a particular point and also at various points in time in a particular zone. Table 1 shows the values of $\overset{*}{w}_i$'s at the three points in time, say 1965, 1970 and 1975. At the same time, Figure 2 illustrates Table 1. Total employment in Osaka Prefecture in those three periods was 3.118 m, 3.601 m and 3.839 m respectively. Osaka city covers Zones 1 to 8 and its population is decreasing even now. The movement of the coefficient $\overset{*}{w}_i$ for ten years shows this fact. In Figure 2 the broken line which illustrates the values of $\overset{*}{w}_i$'s in 1965 lies over two real lines which illustrate those for 1970 and 1975 respectively. On the whole, we could say that the attracting coefficient explains and reflects the real tendencies of residential location. Zones 9 to 18 are located around Osaka city and have developed rapidly and continuously as residential towns. Figure 4 illustrates this. As for these zones, the attracting coefficients also reflect the suburbanisation evident in Figure 2. We may call Zones 1 to 8 a developed region, 9 to 18 a developing region and 19 to 22 an undeveloped region. As a matter of fact, the region covered by 19 to 22 has not yet been fully developed. Figure 2 shows this clearly. The values of $\overset{*}{w}_i$'s for Zones 19 to 22 remained largely unchanged over the ten years from 1965 to 1975. Zone 22 is influenced by the neighbouring city, Wakayama, but over the period 1970 to 1975 Zones 19 to 22 remained unchanged. After all, attracting coefficients explain the tendency of the changing residential locations. Next, we investigate the relation between attracting coefficients and the observed residential locations. This is shown in Figure 3, where the horizontal axis indicates the attracting coefficients $\overset{*}{w}_i$'s in 1975 and the vertical axis the normalised value of L_i in 1975 as

$$\overset{*}{L}_i = L_i \, \frac{22}{\Sigma_i L_i} \quad .$$

The vertical axis also shows the growth rate in L_i between 1970 and 1975, say

$$g_i = \frac{L_i \text{ in } 1975}{L_i \text{ in } 1970} \, .$$

The notations $\overset{*}{L}_i$ (1975) and g_i (1975/1970) in Figure 3 show these variables, respectively. As for the first relation between $\overset{*}{w}_i$'s and $\overset{*}{L}_i$'s, there is the linear relation, as is evident in Figure 3. That is why the attracting coefficient can reflect the tendency of the changing residential locations, as noted above.

As for the $\overset{*}{w}_i$'s and g_i's, on the other hand, no explicit relationship can be found. Therefore, in order to specify the form of $\overset{*}{w}_i$, we must investigate the time series data on $\overset{*}{w}_i$'s in relation to the factors which influence the number of residents in each zone. These factors may be related to the capacity for housing in each zone, say number of houses, rooms or areas. This should be done carefully by factor analysis before the proposed residential location model of Equation (4.4) will give a sound estimate of residential location.

Table 1. Values of $\overset{*}{w}_i$'s for three years (1965, 1970 and 1975)

Zones	1965	1970	1975
1	0.315	0.189	0.140
2	1.022	0.758	0.665
3	0.988	0.783	0.722
4	0.970	0.734	0.618
5	1.094	0.907	0.682
6	1.228	1.125	1.256
7	1.622	1.205	1.061
8	1.460	1.117	1.014
9	0.074	0.104	0.104
10	1.592	1.625	1.685
11	1.380	1.543	1.712
12	0.665	0.930	1.148
13	0.738	1.036	1.281
14	1.209	1.488	1.457
15	1.813	1.766	1.705
16	0.931	1.040	1.103
17	0.631	1.096	1.222
18	1.523	1.616	1.835
19	0.259	0.314	0.327
20	1.393	1.412	1.412
21	0.758	0.726	0.720
22	0.337	0.486	0.485
Total	22	22	22

REFERENCES

[1]   Kondo, K., Estimation of trip demand by chain technique (1977), unpublished Dr.-Eng dissertation at Kyoto University (in Japanese).

[2]   Snickars, F., and Weibull, W., "A minimum information principle: theory and practice" (1977), Regional Science and Urban Economics, Vol 7.

[3]   Murchland, J.D., Applications history and properties of bi- and-multi-proportional models (1977), paper submitted to London School of Economics seminar.

[4]   Murchland, J.D., Convergence of gravity model balancing operations (1977), proceedings of PTRC Summer Annual Meeting, PTRC-P153.

[5]   Lowry, I.S., A model of metropolis (1964), RAND Corporation RM-4035-RC.

[6]   Sasaki, T., Probabilistic models for trip distribution (1968), proceedings of the 4th International Symposium on the theory of traffic flow held in Karlsruhe.

[7]   Wilson, A.G., A statistical theory of spatial distribution models (1967) Transportation Research, Vol 1, No. 3.

[8]   Wilson, A.G., "Generalising the Lowry model" (1971), in Urban and Regional Planning, ed. by A.G. Wilson

APPENDIX

As for the entropy model of the present pattern type, the iteration for convergence was carried out until the following inequalities were satisfied:

$$\left| \frac{\overset{*}{\lambda}_i(2) - \overset{*}{\lambda}_i(1)}{\overset{*}{\lambda}_i(1)} \right| < \varepsilon_E \quad \text{and} \quad \left| \frac{\overset{*}{\mu}_j(2) - \overset{*}{\mu}_j(1)}{\overset{*}{\mu}_j(1)} \right| < \varepsilon_E$$

where ε_E is a criterion for convergence. In the Detroit model, however, the iteration was carried out until the following inequalities were satisfied:

$$\left| F_i^{(1)} - 1 \right| < \varepsilon_D \quad \text{and} \quad \left| G_j^{(1)} - 1 \right| < \varepsilon_D$$

where ε_D is a criterion for convergence. Both iterations reached the same OD table when $\varepsilon_E = 0.001$ and $\varepsilon_D = 0.0001$. The number of iterations for convergence was six in the entropy model and eleven in the Detroit model.

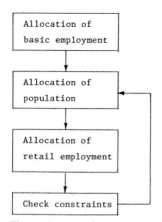

Figure 1. Causal structure of Lowry model

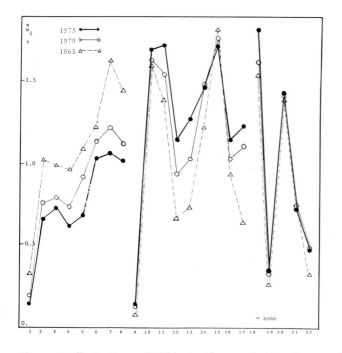

Figure 2. Illustrations of Table 1: Change of attractiveness by year

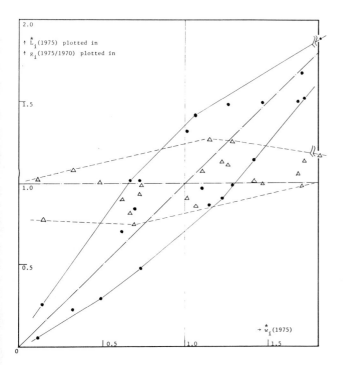

Figure 3. Relations between $\overset{*}{w}_i$ (1975), $\overset{*}{L}_i$ (1975) and g_i (1975/1970)

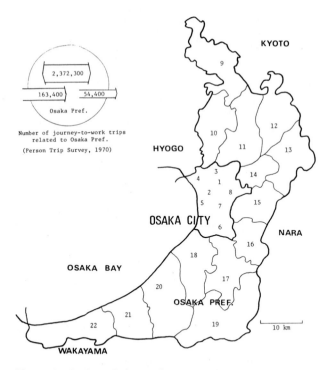

Figure 4. Zoning of the Osaka metropolitan area

Up-to-date training for transportation engineers

8

António José de Castilho   Portugal

SYNOPSIS

In Portugal there are shortcomings in the education and professional training of transportation engineers. The main reasons are the rapid development of the profession in the last few years, the changing role of the transportation engineer, and insuffucient coordination between the different parts of the country's technological structure, right from the planning stage.

The theme of the 1977 congress of the Portuguese Association of Engineers was the training of engineers, and the sub-theme on transportation engineering attracted the greatest number of papers.

INTRODUCTION

In Portugal, subjects related to the changing role of the transportation engineer arouse much interest nowadays. In fact, besides the usual causes, other reasons have contributed to the fact that themes such as "new areas of activity within the scope of transportation engineering", "transportation engineering as a multidisciplinary arena", "cooperation between transportation engineers and urban planners", and "public participation in transportation planning" have reached their present degree of importance in that country.

Engineers, as professionals in the field of transportation, have become increasingly aware of the relevance of these questions, and an action group on transportation engineering has been set up within the Portuguese Association of Engineers. This group has met regularly, and the results of its activities are beginning to appear. One of the questions often raised deals with the new profile of the transportation engineer, with a view to improving the training of engineers in this area. It should be emphasised that training is used in its broader sense, covering both academic education and the concept of continuing education, which seems to be essential today.

TWO STANDPOINTS ON THE TRANSPORTATION ENGINEER'S ROLE

One of the traditional concepts of the transportation engineering is largely concerned with the technologies of each mode of transportation. Some universities follow this concept as they teach these subjects as part of the courses for civil engineering, mechanical engineering and electrical engineering, particularly the first.

Chronologically, it seems possible in a rather schematic manner to define stages in which the emphasis in transportation engineering education was successively laid - first on construction and project; then on traffic and safety; and perhaps last on planning. As a matter of fact, problems and actual needs are likely to bring about adjustments in solutions and progress towards new attitudes as a result of the adaptation and updating of the attitudes prevailing before.

In Portugal it is recognised that there are shortcomings in the education and professional training of engineers. The role of the transportation engineer has gradually been changing in time with the rapid evolution in this field, but some inadequacy may still be detected in the existing management structures. Moreover, there is insufficient coordination between the various bodies which contribute to the technological structure of transportation. These bodies include government departments; enterprises for rail, road, air and water transport; designers and consultants; building enterprises; industry; applied research bodies; and education.

The work of the transportation engineer in each of the sectors mentioned may be chiefly concerned with one of the following activities: design, construction, research and development, exploitation, planning, training, management-administration, or marketing. It seems to be generally accepted that the inadequate coordination, of which harmful results have been detected, begins at the planning stage, for instance between those bodies concerned with the short-term aspect of the project and the execution of the works, and those concerned with action in the medium or long term, for instance applied research, education, and the management of the system itself.

Another aspect that may help in understanding some deficiencies in the technological structure of the transportation sector in Portugal is the unsatisfactory way in which the application of foreign technology has sometimes been resorted to. Instead of improving national technology, this may sometimes weaken it, especially when applied research and education are ignored, thereby hampering or even preventing the up-dating of national technology. Insufficient definition of a task and insufficient subsequent study and discussion may also lead to the same result.

A second concept of transportation engineering might present an opposite view. This view might go from the general to the particular, instead of the opposite. Some hold the view that transportation falls within the scope of production engineering. In this case, the emphasis is laid mainly on system and planning engineering, but to be successfully applied, this requires suitable technological training in each mode of transport, to the extent required for the particular objective. This training, however, has until now not been sufficiently concerned with - for example - the design of the infrastructures for each mode of transportation. A doubt arises - and perhaps this is one of the main shortcomings - about the possibility of production engineers acquiring the minimum of specialised technological training considered indispensable for valid practical decisions.

COOPERATION BETWEEN VARIOUS SPECIALISTS IN RESEARCH

It is generally agreed that regional planning, town planning and transportation planning must be integrated, in order to coordinate data from town development, land use, and national planning and development. At the Laboratório Nacional de Engenharia Civil (LNEC) in Lisbon, a study on the development and application of integrated analytical models of transportation and land use is being undertaken by a transportation engineer working mainly with an urban planner.

At the LNEC there is still an integrated conception of applied research which encompasses comprehensive research plans concerning distinct areas of activity (buildings, hydraulics, etc). One of these is a four-year plan on roads, railways and airports, which essentially consists of the following studies.

Layout. Updating and preparation of automatic computation programmes for the transportation network.

Exploration and earthworks. The study and development of new techniques for geotechnical exploration; studies on the erosion of slopes; stability and distribution of stresses in slopes, in static and dynamic conditions; crossing of alluvial lowlands; and the hydraulic deisgn of drainage works.

Pavements. Design of pavement reinforcement; studies on the non-skidding properties of road pavements and their materials; studies on the fatigue behaviour of road materials, studies on the influence of water on the carrying capacity of road pavements; and adaptation of a set-up for load tests in airports.

Traffic, safety and transportation. Systematisation of accident records and methods for quick detection of high accident locations, particularly on more intensively travelled roads; a study of high accident locations and the influence of road characteristics on them; a study of road safety equipment; a study of paints for road markings, and signs and paints for bridges; a study of accidents in urban areas; and a study of the use and safety of public service vehicles.

Tunnels and underground works. Several studies are in progress.

Engineering studies. Observation and testing of large structures; application of automatic computation to structural analysis and design of bridges; bridge earthquake engineering; and hydraulic problems concerning bridges.

Railways. Study of the stability of railway platforms.

Pollution. Vibrations produced by road and rail traffic and by blasting; and road and air traffic noise (quantification of nuisance factors and forecasting of characteristics).

Dissemination of information on research results. Preparation of recommendations for construction, preparation of specifications; and cooperation in professional training.

THE PRESENT ROLE OF THE TRANSPORTATION ENGINEER

As for the two distinct standpoints on education in transportation engineering, briefly described earlier, the basic aspects they have in common may well be preserved in future, though different approaches will continue to exist, one being more generalist than the other.

In any event, what seems mandatory is the suitable framing of the study subject of which knowledge is considered indispensable today, so that problems can be correctly approached by professionals with either one of the basic educational backgrounds, so that a satisfactory intermediate solution can be arrived at.

In Portugal, a rather small country, there is also the problem of how to provide suitable professional employment for everyone trained in this area, for which specialisation should not be too severe. One of the possibilities would be higher specialisation through postgraduate continuing education.

Some aspects which have been somewhat neglected can assume increasing importance in future. For example, general transportation problems should be viewed from a sociological viewpoint rather than merely from an economic or technical viewpoint. Such a view is backed up by the following statement: "Transport is not concerned with how to move vehicles; but how to transport human beings and goods from origin to destination." From this standpoint, special significance must be given to problems of the environment (such as noise, vibration, air pollution, endangered water resources, and degradation of natural environment) energy, safety and other human problems. There needs to be greater participation by users in decision making in the area of transportation.

BIBLIOGRAPHY

Brandão, F., O ensino da opção de "vias de comunicação" na Faculdade de Engenharia, Lisbon, 1977.

Cameira, A., Que formação deve ter o engenheiro ferroviário?, Lisbon, 1977.

Castilho, A.J. de, A formação de engenheiros nos domínos das vias de comunicação e dos transportes - importância e nível técnico - científico dos estudos referentes a traçado, tráfego, segurança e ambiente, Lisbon, 1977.

Castro, F., Exigências de moderna engenharia ferroviária, Lisbon, 1977.

Costa, E., Formação do engenheiro de estradas, Lisbon, 1977.

Ferreira, T., et al, O caminho de ferro e a formação dos engenheiros, Lisbon, 1977.

Frybourg, M., Les composantes de la modernisation des transports, Enseignmement Supérieur de Transport, Institut de Recherche des Transports, Paris, 1977.

Geraldes, P., Estudos relativos a modelos urbanos e inter-regionais de transportes. Os modelos de procura no processo de planeamento de sistemas de transportes, Laboratório Nacional de Engenharia Civil, Lisbon, 1976.

Gonçalves, E., Os transportes e a formação do engenheiro civil, Lisbon, 1977. Ordem don Engenheiros, Conclusões sobre Engenharia de Transportes, Congress 77, Lisbon 1977.

Laboratório Nacional de Engenharia Civil, Plano de estudos es estradas, caminhos de ferro e aerórdromos, Lisbon, 1977.

Nascimento, U., Dificuldades do LNEC no recrutamento de engenheiros do domínion das vias de comuniçãcao, Lisbon, 1977.

Noortman, H., et al, Panel discussion, World Conference on Transport Research, Rotterdam, 1977.

Pedroso, J., Contributo papa o esboço de uma ajustada formação de engenheiros de transportes, Lisbon, 1977.

Reis, A., et al, A formação de engenheiros no domínio das infraestructuras aeronáuticas e aeroportuárias, Lisbon, 1977.

PART II: TRAFFIC AND TRANSPORTATION PLANNING

Simplified transportation system planning   9

H. Ayad and J. C. Oppenlander   United States

SYNOPSIS

Traffic assignment is the process of allocating trip interchanges to a transportation system. This operation is performed to reproduce both present and future traffic flow problems, evaluate proposed plans or optimise network flows. Although the major use of traffic assignments in urban transportation studies is to evaluate proposed plans, no analytical procedure has been available to quantify the adequacy of a transportation system.

The purpose of this study was to develop a rational concept for the design and evaluation of urban transportation systems. The concept is based on the premise that the adequacy of a plan is described as the degree to which its design features satisfy the objectives of the community in transportation services. From the standpoint of transportation, these objectives are defined as the attainment of selected levels of service for trip interchanges between pairs of urban zones. A plan is considered adequate when the transportation facilites accommodate to a reasonable degree the traffic movements at these desired service levels.

A desire-assignment procedure, which is identified as the Simplified Proportional Assignment Technique (SPAT), is employed to determine the nature, magnitude, and location of deficiencies in a transportation system. Trip interchanges are assigned on a proportional basis to acceptable routes which satisfy the pre-set levels of service. The assignment technique leads to the detection of link and zonal deficiencies. Link deficiencies occur when the assigned volumes on street segments exceed the service volumes of these sections. If there is no acceptable route between a zonal pair, then there is a zonal deficiency, and trip interchanges can be accommodated only by improvements in the system. Link and zonal deficiencies define the nature and extent of the improvements needed to make a transportation plan adequate.

This technique of transportation planning permits a comparative appraisal of the costs of providing urban transportation systems for various specified levels of service. As a result, the application of this procedure allows the selection of a transportation system that provides a balance between community goals, as expressed by levels of service, and the associated costs involved in eliminating deficiencies in the transportation network. This cost-effective approach can be applied to any system of transportation, such as motor vehicles, mass transit conveyances, bicycles, pedestrians or goods shipments. A transportation plan for a small community was developed by this simplified transportation planning process.

INTRODUCTION

The allocation of travel movements to a transportation system is known as traffic assignment. The interchanges between selected origins and destinations can be person or vehicle trips for a single or several modes of travel. The assignment of trip interchanges to a transportation system is performed for the following purposes: reproduction of travel patterns, evaluation of proposed plans, and optimisation of network flows.

In the reproduction of travel patterns, trips are allocated to the transportation system so that the assigned volumes are similar to the actual volumes accommodated on the links of the network under the existing or projected conditions. The procedure may involve the allocation of the existing or a projected set of trip interchanges to the existing or a proposed system. The process is employed to study the changes in flow patterns produced by either a change in the operational controls or by the addition of new facilities to a system.

Traffic assignment is also used to evaluate proposed plans and to detect deficiencies in transportation systems. When the technique is used for this purpose, trips are allocated only to desired routes. Between an origin and a destination, there may be one or more desired routes for the assignment of trip interchanges. Projected travel movements are allocated to either existing or proposed plans. Comparison of assigned volumes with available service volumes is a measure of the possible deficiencies in the various sections of the system for the design period under consideration.

A third application of traffic assignment is the determination of the optimum use of a transportation network. Trip interchanges are allocated to the network to optimise one or more chosen travel parameters, such as the total cost of travel or the total time spent by drivers on the network. From this assignment, appropriate traffic engineering control measures are selected to regulate the movement of traffic in the real system in accordance with the conditions producing optimum flow.

The major problem in the attainment of the "best" transportation plan for a community stems from the present approaches to the development of these plans. At present there are no exact methods for developing elements and for testing the adequacy of a transportation system in carrying a given set of trip interchanges and in meeting the study objectives. Plans are developed on the basis of rough methods of estimating traffic flows, such as the all-or-nothing non-capacitated assignment. In addition, the capacity-restraint models neither reflect travel desires nor point to system deficiencies. Their application to a developed plan does not ensure that the plan adequately meets the study objectives. It is possible to accept badly planned networks with balanced volumes on them by using capacity-restraint models.

The purpose of this research investigation was to develop a rational concept for the design and the evaluation of transportation systems.[1] The concept was designed to permit the analysis of any transportation system by quantifying the degree to which its facilities satisfy the study objectives and the travel desires of the community. This new concept, identified as the Simplified Proportional Assignment Techinque (SPAT), was based on the premise that the community objectives, as related to transportation services, are the attainment of particular levels of service between origin-destination combinations within the study area. These levels of service may differ from one community to another and may vary over the years for the same community. Proper service levels are selected to reflect the desires of the people for a transportation system which must be restricted by financial and technological limitations.

The application of this procedure permits the selection of a transportation system that provides a balance between community goals, as expressed by levels of service, and the associated costs involved in eliminating deficiencies in the transportation network. This cost-effectiveness approach can be applied to any system of transportation, such as motor vehicles, mass transit conveyances, bicycles, pedestrians, or goods shipments.

CONCEPTUAL MODEL FOR NETWORK EVALUATION

The adequacy of a proposed transportation plan is evaluated by studying the degree to which the design satisfied the community objectives. In this investigation, the transportation goals of an urban centre are represented by the attainment of specific levels of service for chosen trip interchanges in the area. Cost and technology limit the feasible range of these service levels and make them time and community-dependent criteria. Standards acceptable to a particular community may be rejected by another urban centre. Similarly, the interpretation of an acceptable level of service may vary over the years in a particular community.

Complete flexibility in the development of a plan is achieved by the use of service levels to describe the community objectives as related to transportation. This approach permits total, partial or no improvements in the existing levels of operation in a system. An existing network may be adequate for the future if the community is willing to accept reduced levels of service. Conversely, an expensive plan based on a high performance is justifiable as long as the people desire it and are willing to pay for its implementation.

The level of service concept is useful in the selection of preferential improvements in a transportation system. Selected trip interchanges may be given higher levels of service than other traffic movements. The resulting plan would favour these preferred trips by providing them with enhanced travel qualities.

Basis of the assignment procedure. Between an origin and destination pair of urban zones, there are routes or corridors of travel to expedite the flow of traffic within established levels of service. The number of these acceptable routings depends on the characteristics of the transportation system and the chosen service level. For some trip interchanges, there may be no acceptable route, then there are zonal deficiencies for the selected service levels. On the other hand, other inter-zonal transfers may have one or more acceptable routes.

Traffic can move on routes at chosen levels of service as long as the assigned traffic volumes do not exceed specific upper limits which are referred to as the service volumes for the various links of the system. When this volume is surpassed on a link, the actual link service level drops to a value below the acceptable limit. Service volume and travel speed for a network selection are defined by a selected level of service to establish the desired quality of traffic flow on that link. The choice of the proper qualities of traffic flow on the network is a decision made by the planning team to reflect the desires and the financial limitation of the community in respect of transportation services.

Establishment of the inter-zonal levels of service and the qualities of traffic flow on the links of a system set a limit to the number of available routes and acceptable links. When improved inter-zonal service levels are desired, the number of acceptable routes is reduced. Similarly, the choice of a better quality of traffic flow on the transportation routes often results in assigned link loadings exceeding the selected service volumes.

The acceptable routes thus delimit the network available to accommodate the projected traffic interchanges within the limits of the study objectives. After the acceptable routes have been developed according to the specified levels of service, the relative attractiveness of these routes is determined for the movement of traffic between selected origin-destination combinations. Trip interchanges are then assigned to these routes on a proportional basis in accordance with their relative attractiveness. This form of traffic assigment reproduces the travel desires as dictated by the selected levels of service for the various trip interchanges.

The determination of acceptable routes between an origin and a destination and the subsequent allocation of trip interchanges to these routes are not limited in application to vehicular movements. An acceptable path between a zonal pair may be obtained by use of mass transit facilities, and the routes to which the trip interchanges are assigned can include single-mode or multi-mode connections. The trips are again assigned on a proportional basis in accordance with the relative attractiveness of the routes and the modes. The attractiveness indices express the relative desirability of particular modes of travel and specific routes within these modal groups.

System evaluation. The application of the above assignment procedure may result in one or more of three possible outcomes - zonal deficiencies, link deficiencies and no deficiencies. Zonal deficiencies occur when there are no acceptable routes to move traffic between chosen zones at specified levels of service. No inter-zonal transfers between these two zones can be accommodated by the considered system and can therefore only be handled by improvements or additions to the available facilities.

The allocation of trips to the acceptable routes between origin-destination combinations may result in the overloading of some links. Overloading of a link occurs when the assigned number of trips exceeds the designated service volume for that link. A link deficiency is the difference between the assigned trips and the service volume on a particular link. In the event of link deficiencies, selective improvements on the transportation system are needed at or near the overloaded links to accommodate the excess traffic.

Finally, the assignment of trip interchanges to the transportation system may result in no significant zonal or link deficiencies for the entire urban area. Because non-deficient links are able to carry the assigned volumes of traffic at the prescribed levels of service, no improvements are required on these links.

Depending on the nature and magnitude of the deficiencies, a plan is developed to accommodate the excess traffic. The ability of a plan to meet the desired objectives is then tested by reassigning the set of trip interchanges for the design year to the proposed network. The adequacy of a transportation system is confirmed when all inter-zonal transfers are accommodated without significant zonal and link deficiencies.

Simplified Proportional Assignment Technique (SPAT). This process of transportation system design and evaluation is identified as the Simplified Proportional Assignment Technique (SPAT). The concept underlying the assignment technique is based on the premise that a transportation system provides selected levels of service for particular trip interchanges. A plan is evaluated by considering the degree to which these travel objectives are satisfied.

A diagrammatic representation of SPAT is given in Figure 1 (page 89). The process, which may be applied for the evaluation of any transportation plan, includes the following operations.

□ Appropriate levels of service are selected for various trip interchanges within the study area. These service levels are predicated in the study objectives as related to the transportation requirements and the financial limitations of the community.

□ The desirable qualities of traffic flow in the various components or links of the transportation system are determined for the selected levels of service and the characteristics of the street or transit line sections. This step establishes the link speeds and the service volumes that different section of the system are able to accommodate at various levels of service.

□ All acceptable routes are determined that connect chosen pairs of urban zones. Acceptable routes include only corridors of travel capable of moving traffic between an origin and a destination at the chosen level of service and quality of traffic flow. One or more modes of travel may be included in the search for these routes.

□ Trip interchanges are then assigned to the travel corridors on a proportional basis reflecting the relative attractiveness of these corridors.

□ Any deficiencies in the system are zonal or link deficiencies. A zonal deficiency results when there is no acceptable route to which a given set of trip interchanges may be assigned. A section that is assigned more trips than its service volume has a link deficiency. The proposed improvements on the system are designed to accommodate the excess traffic associated with both link and zonal deficiencies.

EVALUATION OF THE CONCEPT PARAMETERS

The various components of SPAT were evaluated to permit the application of the proposed concept to urban transportation studies. There are two types of parameters employed in this planning procedure. The first refers to the items requiring quantification by the planning team. Because these variables are closely related to community objectives and policies, their evaluation depends on the goals of the community for which the study is performed. The other parameters are related to the computational aspects of the assignment technique. Measures for three variables can be established and used for any transportation study.

Definition of study objectives. A plan is evaluated by determining the degree to which the design features satisfy the stated objectives. The community objectives in relation to the desired transportation services are expressed as desired levels of service for trip interchanges between pairs of urban zones. When SPAT is used to evaluate an existing or proposed transportation system, these service levels are considered in the search for acceptable routes. An acceptable path is determined on the basis of satisfying a chosen inter-zonal level of service. Numerous attempts were made to identify the factors associated with a level of service on an urban facility. Although the driver's motivation on a particular trip encompasses a wide spectrum of possibilities, four parameters are considered indicative of this motivation. These factors are travel time, operating cost, safety, and driving consideration.

In the present version of SPAT, acceptable routes are selected on the basis of travel time only. However, the other variables that define a level of service are considered in the proportional assignment of trips to these routes. The total travel time is not a realistic parameter to define the study objectives because inter-zonal distances change for different zonal pairs. For a systematic computational procedure, overall travel speed is a more appropriate parameter than total travel time. This change in variable does not affect the determination of acceptable routes; instead, it permits the expression of the study objectives in more convenient measures. In actual application, the acceptable overall speeds to be used in SPAT may vary with communities. The appropriate values depend, among other things, on the economic base and the stage of development of the community and on the purposes of the trip interchanges.

Selection of the quality of traffic flow. Another evaluation parameter which may quantitatively vary from one study to another is the desired qualities of traffic flow on various segments of the transportation system. When this value is coupled with the inter-zonal levels of service, the qualities of traffic movement establish the nature and complexity of an adequate plan. An expensive plan is one in which high inter-zonal levels of service and superior qualities of traffic flow are selected. On the other hand, little or no improvement may be needed in an existing system if sufficiently low standards are adopted for both the inter-zonal levels of service and the qualities of movement.

The quality of traffic flow recommended for use in this evaluation of transportation systems corresponds to the level of Service C as specified in the Highway capacity manual.[2] This link level of service indicates stable flow conditions in which many drivers experience some restrictions in the traffic stream. If superior traffic flow qualities are desired, or inferior travelling levels of service are acceptable, Levels B or D, respectively, may be used.

Acceptable routes in a network are determined on the basis of impedance factors assigned to the various links in accordance with the selected levels of service. Because the assignment technique is designed to reflect travel desires and not to reproduce flow patterns, the impedances used in route determination are not revised after allocating the trip interchanges to the system.

Determination of acceptable routes. SPAT is based on the premise that drivers select from two or more acceptable routes in their movements between specified origin-destination combinations. Alternative route determination in this method of system evaluation refers to the selection of distinct corridors of travel within the study area. A minor difference between two paths does not justify considering them as separate corridors for traffic movement. From the standpoint of transportation planning, corridors of travel desires describe adequately the channels of traffic flow.

If distinct routes of travel are to be developed for selected zonal interchanges, considerable portions of their lengths should represent separate travel corridors and they may share only limited segments of the street system, and then only in the vicinity of the terminals of the trip and over certain control sections in the transportation system. Control sections generally include tunnels and bridges which afford the only access between various districts in the study area.

Based on these features of the corridors of travel, acceptable routes between zonal pairs are determined in the following manner.

□ The minimum-time path joining a chosen pair of zonal centroids is determined by an appropriate algorithm.

□ If this minimum-time path satisfies the pre-set level of service restriction, a central percentage of this path is removed from the network description. A 7 per cent overlap at either end of alternative routes is stipulated as permissible in the present computer version of SPAT. Only control sections on the route, such as bridges and tunnels, are exceptions to this rule.

□ The minimum-time algorithm is again employed to find the second-best route for the reduced network, and the travel time on this route is compared with the pre-set service level.

□ A central percentage of this second-best route is now removed from the network description if this routing is acceptable in terms of the level of service criteria.

□ The process of minimum-time path search and the removal of the central section of the determined route is repeated until the minimum-time route over a particular network description does not fulfil the service level requirements. In addition to satisfying the overall travel speed as the specified inter-zonal level of service, the acceptable routes in SPAT must provide travel times and route lengths not exceeding 1.75 of the corresponding values on the initial minimum-time path for each zonal pair. The final set of acceptable routes comprises all feasible corridors of travel that accommodate traffic between the prescribed zones at the established levels of service.

Proportioning trips among alternative routes. When there are several acceptable paths between an origin-destination pair of zones, the zonal interchanges are proportioned among these available travel routes. The basis of proportionality is predicated in the relative attractiveness of these routes. Parameters used to proportion traffic to alternative routes include travel time, trip distance and driving comfort. The purpose of a trip and the characteristics of the drivers determine to a large degree the relative weights of these three factors in proportioning trips among alternative routings.

Driving comfort is closely related to the nature of the facilities used for a particular trip. The tension experienced on a route is greatly affected by traffic interferences, such as signalised intersections, parking manoeuvres, and access to adjacent property. Nodes are incorporated in the network description at intersections of streets and at major points of access to abutting property. Because nodes represent locations of traffic interference, the number of nodes on a route may be related to the tension experienced by the drivers on that route.

The procedure in SPAT for proportioning traffic among several acceptable routes includes the following sequence of operations.

□ An attractiveness index is calculated for each acceptable route in terms of the following equation:

$$F_i = \frac{1}{T_i^a} \times \frac{1}{D_i^b} \times \frac{1}{N_i^c}$$

where F_i = attractiveness index of route "i"; T_i = time travel on route "i"; D_i = total distance of route "i"; N_i = number of nodes on route "i"; a = exponent of travel time; b = exponent of total distance; and c = exponent of number of nodes.

☐ The trips to be assigned to each route are obtained from the following relationship:

$$L_i = L \times \frac{F_i}{F_1 + F_2 + \ldots + F_n}$$

where L_i = trips assigned to route "i"; L = total distributed trips to be assigned to the acceptable routes for a zonal interchange; and n = number of acceptable routes between a pair of zones.

The values of exponents a, b, and c are dependent on the trip purpose. A sensitivity analysis was performed on the proportionality factor for varying values of the exponents of travel time, total distance, and number of nodes. The resulting attractiveness indices varied only by 5 to 10 per cent for various combinations of exponent values, ranging from the square root to the second power. In order to demonstrate the application of the new concept, exponents for these three variables were taken as 1.0 in the application of SPAT.

APPLICATION OF SPAT

To demonstrate this process for the planning of transportation systems, SPAT was applied to the projected trip interchanges for the city of Monroe, North Carolina. The design year of 1990 was selected, when the population of this community is projected to be 17,000. The data concerning the projected traffic movements on the network of this urban centre were obtained from a study report prepared by the North Carolina State Highway Commission.[3] Some modifications of street layout and zonal subdivisions were introduced, however, to obtain a clearer and simpler representation of the land-use arrangement and the transportation system.

The evaluation of proposed plans by SPAT involves the selection of decision-making parameters relating to levels of service for trip making in the community. These variables include the inter-zonal levels of service and the qualities of traffic flow for the links in a transportation network. The planning objectives as related to urban transporation were selected to provide this community with the inter-zonal levels of service shown in Table 1 for the various categories of trip interchanges. This combination describes acceptable inter-zonal travel speeds ranging from 15 mph for intra-core movements to 30 mph for through trips. The desired quality of traffic flow on the various links of the street network was specified as level of Service C, as defined in the Highway capacity manual.[2] The travel speeds and service volumes for the network links were established from the information provided in Table 2.

Peak-hour interchanges were calculated as 10 per cent of the average daily traffic as forecast for the future trip table. This technique of transportation system evaluation was first applied to the existing transportation facilities for the origin and destination zones shown in Figure 2 and the coded network described in Figure 3. A computer programme was prepared for the convenient and efficient application of SPAT. SPAT showed there were both link and zonal deficiencies in the present network. These deficiencies are described in Tables 3 and 4 under the heading "Existing System". A total of 19 trip interchange combinations and zonal deficiencies, and 4169 unaccommodated vehicles constituted the link deficiencies in the system.

The nature and the location of the improvements needed were then determined by analysing the deficiencies shown up by the evaluation process. To demonstrate the effects of the adopted improvements on the adequacy of the transportation system, the proposed facilities that would accommodate the excess traffic were considered in three successive stages for this illustrative application of SPAT. These stage improvements are shown in Figure 4.

The first stage of the recommended improvements included the construction of an expressway segment between nodes 49 and 72 and a street section between nodes 47 and 49. The SPAT computer programme was employed again with the first proposed transportation plan which featured the inclusion of the first set of improvements in the existing transportation system. The zonal and link deficiencies shown in the new assignment process are detailed in Tables 3 and 4 under the heading "First stage improvements". The first stage of improvements reduced the zonal deficiencies from 19 to 9 and decreased the link deficiencies from 4169 to 3173 unaccommodated vehicles.

A second set of improvements was then proposed to eliminate the remaining deficiencies. These modifications included interchanges at nodes 50 and 84 and a widening of the street segment between nodes 76 and 77 to four lanes. The computer programme was applied once more to the expanded network comprising both existing facilities and proposed improvements. The deficiencies are shown in Tables 3 and 4 under the heading "Second stage improvements". Although the number of zonal deficiencies was not affected by the second set of improvements, the link deficiencies were reduced from 3173 to 1649 unaccommodated vehicles.

More new facilities were adopted, and the system evaluation technique was also employed for the newly proposed plan. The third set of improvements included three street segments (in Figure 4) providing direct connections between the zonal pairs 76 and 78, 71 and 78, and 73 and 80. The deficiencies shown up by the last assignment are presented in the column "Third stage improvements" in Tables 3 and 4. These inadequacies included one zonal deficiency and 1163 unaccommodated vehicles on seven deficient links. The link deficiencies ranged in value from 68 to 360 vehicles per link for the peak-hour trip interchange loadings.

Although the three stages of improvements did not result in full attainment of the study objectives, some congestion on a few links was considered acceptable. The improvements needed to account for the resulting inadequacies (one zonal and seven links) did not appear to justify further investment to eliminate these deficiencies. In transportation studies performed by operating agencies, minor localised deficiencies may be tolerated at lower levels of service when these deficiencies require unjustifiable investments to accommodate the excess traffic on the system. In addition, the degree of completeness achieved in this build-up process of transportation system planning is probably well within the precision limits obtainable for the forecast trip interchange values. On the other hand, this process can be repeated with additional network improvements until all zonal and link deficiencies are eliminated. Any improvement that is not cost-effective can, of course, be identified and eliminated in this technique of transportation system planning.

The Monroe Street network, which was used in demonstrating the application of SPAT, had 37 zonal centroids, 88 nodes and 264 links. The average computer time to execute a complete assignment operation for this network was six minutes for the present version of SPAT.

SUMMARY AND CONCLUSIONS

This research investigation was concerned with the development of a new concept traffic assignment for the evaluation of urban transportation systems. The follow operations comprise the main features of the technique.

□ The study objectives as related to transportation in a community are expressed as the attainment of specific levels of service between pairs of urban zones within the study area.

□ All routes that can move traffic at these prescribed levels of service are determined, and trip interchanges are assigned to these routes on a proportional basis. The evaluation procedure may result in the detection of zonal or link deficiencies. A zonal deficiency arises when there is no route that can move traffic at the established inter-zonal service level, and a link deficiency occurs when the desirability to use a street segment exceeds the ability of this segment to provide a specified quality of traffic flow.

□ This procedure is applicable to either an existing or a proposed transportation system. The adequacy of a proposed plan is confirmed when the transportation facilities are able to accommodate a given set of trip interchanges at the desired levels of operation without either zonal or link deficiencies.

SPAT is a practical and reliable procedure for network evaluation. The use of the technique in urban transportation studies permits the quantitative determination of the adequacies of proposed plans. Becaused SPAT is used to evaluate a transportation plan by direct reference to the study objectives, differences among communities in desires and resource limitations are incorporated in the procedure. SPAT provides a rational technique that employs traffic assignment to quantify the adequacies and deficiencies of urban transportation systems.

REFERENCES

[1] Ayad, H., System evaluation by the Simplified Proportional Assignment Technique (1967), Thesis submitted to Purdue University for the degree of Doctor of Philosophy.
[2] Highway capacity manual (1965), Highway Research Board, Special Report No 87.
[3] Monroe thoroughfare plan (1966), North Carolina State Highway Commission.

Table 1. Inter-zonal level of service used in Monroe plan evaluation

Trip type	Corresponding zones	Level of service (mph)
Intra-core	1,2,3	15
Intra-city	4 through 11	20
To-core	trips to 1,2,3	20
Through	26 through 37	30
All others		25

Table 2. Speeds and volume-to-capacity ratio on links
for designated level of service in Monroe plan evaluation

Link level of service	C
	Speed (mph)
Core	20
City	25
Fringe	30
Volume to capacity ratio	0.8

Table 3. Link deficiencies in vehicles per hour

Link	Existing system	First stage improvements	Second stage improvements	Third stage improvements
31 - 88	288	290	-	-
39 - 43	314	268	268	280
42 - 43	70	-	-	-
43 - 44	150	-	-	-
46 - 59	32	-	-	-
48 - 50	136	135	-	-
49 - 56	536	358	360	360
50 - 52	65	71	-	-
52 - 53	162	154	164	140
53 - 54	68	69	71	68
54 - 74	82	71	72	71
56 - 61	210	131	132	92
61 - 67	91	-	-	-
63 - 69	172	80	80	-
66 - 67	130	-	-	-
67 - 68	230	-	-	-
69 - 70	296	186	186	-
69 - 73	37	16	16	-
72 - 78	-	96	100	-
75 - 76	28	50	50	-
76 - 77	222	242	-	-
78 - 86	64	146	150	152
83 - 84	314	338	-	-
84 - 85	192	190	-	-
84 - 88	280	282	-	-
Total	4169	3173	1649	1163

Table 4. Zonal deficiencies

Link	Existing system	First stage improvements	Second stage improvements	Third stage improvements
3 - 7	x			
4 - 10	x	x	x	
4 - 15	x	x	x	
4 - 18	x			
7 - 12	x			
7 - 13	x			
8 - 12	x			
8 - 13	x			
8 - 14	x			
9 - 14	x	x	x	
10 - 14	x	x	x	
11 - 15	x	x	x	
11 - 16	x			
27 - 33	x			
30 - 34	x	x	x	
31 - 34	x	x	x	
31 - 35	x	x	x	x
33 - 35	x	x	x	
33 - 37	x			

x = Existence of a zonal deficiency

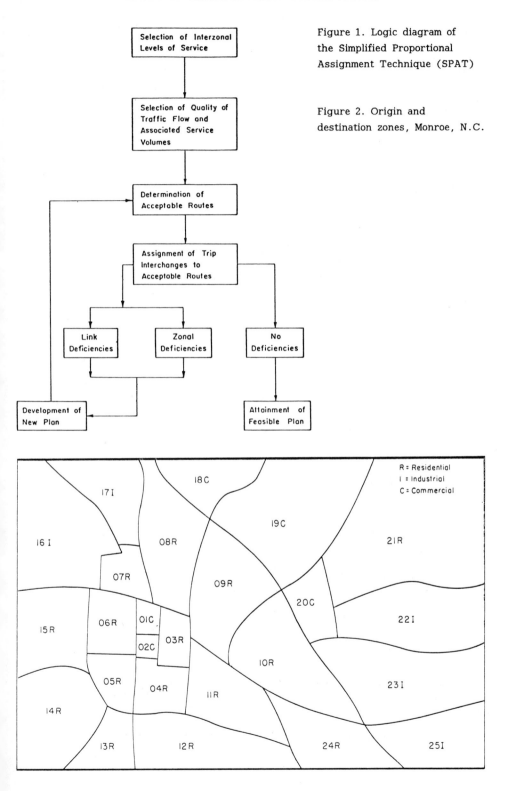

Figure 1. Logic diagram of the Simplified Proportional Assignment Technique (SPAT)

Figure 2. Origin and destination zones, Monroe, N.C.

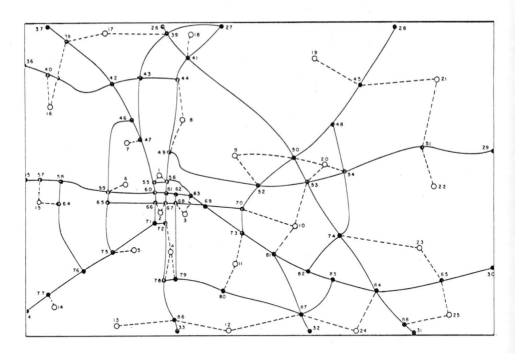

Figure 3. Coded network, Monroe, N.C.

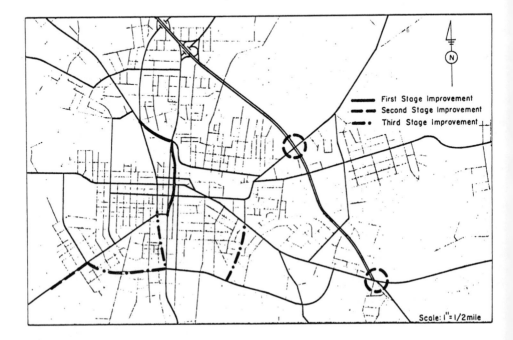

Figure 4. Stage improvements, Monroe, N.C.

Policy-sensitive and policy-responsive transportation planning

10

Moshe Ben-Akiva and others   United States

SYNOPSIS

This paper reviews recent developments and new approaches representing significant improvements and broadening of capabilities in the methods and procedures available for transportation planning. Foremost among the theoretical and methodological developments has been the application of statistically estimated disaggregate models for predicting transportation demand. This has led to the development of more realistic model systems that are policy-sensitive and produce forecasts in agreement with observed changes. Other advances, such as improved techniques for the equilibration of transportation demand and network performance, have also proved important, but none quite so fundamental in their implications for alternative analysis approaches which are policy-sensitive and responsibe to changing analysis needs and accelerating decision processes.

   The development of simplified procedures that require less data, time and money to prepare and use also benefited directly from increasing computational capabilities - the development of powerful and economic pocket calculators, especially programmable and card-programmable ones, and mini-computers. The paper presents the basic characteristics of the newly developed procedures. The most notable are: efficient use of available data sources with small-scale and inexpensive surveys; adaptability of the analysis procedures to different levels of geographical detail; and model simplifications achieved primarily by reduced input data requirements without a loss in policy sensitivity. The paper concludes with examples of analysis procedures that have been implemented and applied in both computer and pocket calculator environments.

INTRODUCTION

During the third quarter of this century urban transportation systems analysis focused on the development of models to describe the urban travel patterns with concepts and techniques borrowed from economics (demand, supply, etc.), civil engineering and operations research. Using these disciplines for a theoretical background and relying on the newly available computational capabilities, which allowed the analysis of immense data sets, researchers assume that urban travel patterns could be forecast for a planning horizon of 15 or 20 years. Contributing to this perspective was the fact that until the late 1960s, solutions to transportation problems were seen mainly as construction options, i.e. increasing network capacity. This type of solution was gradually discarded as the social, economic and environemtal implications drew wide public and political interest. A new range of transportation options was needed, and tools to evaluate them were absent.

Perhaps the turning point in this era was Williamsburg, Virginia, in December 1972. There, at a meeting of urban transportation analysts, responsible policy officials of the US Department of Transportation challenged the urban transportation planning profession. They warned that the elaborate computer-based methodology which those professionals had been engaged in developing and applying for some fifteen years was not at all responsive to the issues which were then politically relevant in urban transportation decision making at local, state and federal levels.[1][2] Moreover, on the basis of the available models, transportation systems analysis was seen as a very expensive planning phase requiring large data collection and set-up efforts, dependent on a large computer.

The disaggregate choice models which were seen in Williamsburg as emerging techniques have been developed into a flexible forecasting approach which addresses the shortcomings of the conventional models. The development of network performance analysis techniques and equilibration techniques has also been improved since, but none of these developments has been as significant for the planning capabilities as the disaggregate travel demand models.

The availability of such models is appropriate for today's planning environment: planning funds are limited and there is a growing need for responsiveness in planning, partially owing to public involvement in the process. Of major importance is the sensitivity of these models to current policy issues.

The purpose of this paper is to provide an overview of recent research and application of transportation planning procedures and demonstrate its key feature: the sensitivity of these procedures to the policies and options which the planner must face today.

BASIC CONCEPTS AND TECHNIQUES

The theoretical basis for the disaggregate choice model is a theory of consumer behaviour, which assumes that any trip is the result of a selection made by an individual traveller from a set of available alternatives. The demand for travel is viewed as a process arising directly from individual decision makers' choices.

In contrast to conventional consumer demand models, the choices made in travel-related decisions are discrete and qualitative. Therefore, the approach of disaggregate travel modelling is based on qualitative choice models.[3] The individual decision-making unit (an individual or a household) is faced with a set of feasible alternative (travel) options from which one is to be selected. Denote the choice set of individual t as:

$$C_t = \{1, 2, \ldots, i, \ldots, J_t\},$$

where J_t is the number of feasible alternatives.

The choice process is analytically modelled using the concept of utility maximisation. Let U_{it} be the utility of alternative i to individual t. Alternative i will be chosen if

$$U_{it} \geq U_{jt}, \quad j = 1, 2, \ldots, J_t.$$

For predictive purposes the relative values of the utilities must be related to observed variables as follows:

$$U_{it} = V(Z_{it}, S_t) + \varepsilon_{it},$$

where Z_{it} = a vector of attributes of alternative i faced by individual t; S_t = a vector of socio-economic characteristics of individual t; and ε_{it} = an unobserved random component of the utility of alternative i to individual t.

The random component in the utility function may be the result of omitted unobservable variables, measurement errors or other errors in the specification of the utility function.[4] Owing to the existence of the random error term, only probabilities of choice can be predicted as follows:

$$P(i|C_t) = P \{ U_{it} \geq U_{jt}, \; j = 1, \; 2, \ldots, J_t\}$$
$$= P \{ \varepsilon_{jt} - \varepsilon_{it} \leq V(Z_{it}, \; S_t) - V(Z_{jt}, \; S_t), \; j = 1, \; 2, \ldots, J_t\}.$$

where $P(i|C_t)$ = the probability that alternative i will be chosen from the set of alternatives facing individual t.

A probabilistic choice model is derived by assuming a specific joint distribution of the random utilities $(\varepsilon_{1t}, \varepsilon_{2t}, \ldots, \varepsilon_{jt})$ for the given values of the observable variables. Most common to date is the use of the multinomial logit (MNL) form which assumes that the random errors are independently and identically Gumbel-distributed.[3] The general form of the MNL model is:

$$P(i|C_t) = \frac{e^{V(Z_{it}, S_t)}}{\sum\limits_{j=1}^{J_t} e^{V(Z_{jt}, S_t)}}$$

The parameters of the systematic utilities $V(Z_{it}, S_t)$ are estimated on the basis of data collected from a random sample which includes information on the available set of alternatives (C_t); the chosen alternative, which is assumed to be the maximum of all the utilities of the available alternatives; the characteristics of the alternatives (Z_{it}); and socio-economic characteristics of the individual (S_t). The most frequently used statistical estimation method is maximum likelihood with linear in the parameters utility functions. A number of efficient computer packages are at present available for estimating logit models.

The disaggregate choice models can be applied to the entire set of travel-related decisions and any specific model is explicitly structured to reflect the choice set analysed. The range of choices for which models have already been suggested varies from long-term decisions, such as employment or residential location and automobile ownership, to short-term decisions, such as frequency, destination, mode, time of day and route choice for non-work trips.

Disaggregate survey data is used directly for the statistical estimation of the models. This fully exploits the information available from a given survey data set. In contrast, aggregate modelling approaches lose a great deal of the variability inherent in existing data by grouping observations at the zonal level. Disaggregate estimation also reduces the potential of biases in estimated model coefficients, owing to the existence of the simultaneous link from travel demand to level-of-service attributes.

The efficiency of data utilisation in the application of disaggregate demand models is further shown by the range of alternative sources of data for prediction. In view of the high cost of data collection, emphasis was given to developing forecasting methods based on readily available data sources. Only rarely is a home interview survey available at the point of time or location in which the planner faces a prediction problem. Some of the suggested methods applied and validated will be discussed below.

Another advantage of disaggregate travel demand models is the significant independence of location-specific variables. The potential transferability of these models among urban areas has been shown by Atherton and Ben-Akiva.[5] The coefficients estimated in widely different urban areas are remarkably similar, even when the mean values of the variables are significantly different. To account for the differences which do exist between environments and which are primarily manifested in the constant term of the utility function, Atherton and Ben-Akiva have suggested a number of methods of updating. These methods range in complexity and cost and include some very simple and inexpensive methods, as will be demonstrated below. The significance of the transferability of the disaggregate models cannot be underestimated: it means that planners can use readily available models which have been estimated elsewhere and, with only minor adjustments, apply them to the problems they are faced with.

Using disaggregate models for aggregate prediction requires the employment of explicit aggregation procedures. Since detailed data on the prediction population is not available, aggregation must be based on limited information on the distributions of socio-economic characteristics and attributes of the alternatives. Several aggregation procedures are evaluated in Koppelman,[6] and some applications of aggregation procedures will be briefly described below.

Analysis of the network performance and equilibrium conditions are the complementary stages of the forecasting procedures. The range of methods available for this analysis vary in the level of analysis from full representation of the network, through various degrees of abstract representation of the network to the analysis of performance on a single link. The appropriate method depends on the necessary levels of accuracy of performance prediction at any specific location and on the cost of the analysis. It should be noted, however, that the most important outcome of the analysis of performance is the level of service offered by the network which is an input of the demand model.

The available equilibration procedures also range in complexity and cost, but generally share a similar iterative structure. A demand model is applied to predict the demand user under an estimated level of service. The predicted demand is then assigned to the network or link and the resultant level of service is then used to calculate the new input of the demand model. The iterative procedure is performed until convergence criteria are satisfied.

ALTERNATIVE PROCEDURES AND APPLICATIONS

Applications in policies and planning studies. The disaggregate models can be applied to aggregate forecasting in a variety of ways, ranging in level of detail and data requirements from detailed sub-area analysis and conventional network-based simulations to a highly aggregate sketch-planning procedure based on a small number of market segments. Some of the recent applications in policy and planning studies are summarised below to demonstrate the models' wide range of applicability and to emphasise the fact that disaggregate models are practical analysis tools. The models have been used for:

□ A policy study on the effects of alternative programmes of incentives for car-pools (shared use of cars for work trips). Washington, DC, and Birmingham, Alabama, were used as prototype cities.[7]

□ Planning studies of car-restricted zones. The models were used to predict the effects of various car-restricted zone concepts in selected cities.[8]

□ A planning study of anticipated guide-way transit strategies for Milwaukee. The models were used in conjunction with an urban transportation planning system in both sketch-planning and detailed network analysis approaches.

□ A planning study of a people mover system for internal circulation within the Los Angeles central business district. Models have been developed for predicting choice of parking lot and egress mode for peak-period trips (travel from parking to destination), for arrivals by auto and transit, and for noon-hour trips, frequency destination and mode of within-CBD trips (modes including walking, minibus and people-mover systems).[9]

Many other applications, primarily of mode-choice models, have been made. See, for example, the review of several applications given by Spear.[10] Thus, disaggregate methods are being used in a variety of practical policy and planning applications.

The range of analysis contexts. As the examples indicate, a very wide range of options can be analysed by using disaggregate demand models integrated with some performance and equilibration procedures. This range can be classified among several dimensions, such as level of analysis, spatial detail, time horizon, and types of options.

An analysis can be performed at a very detailed level, predicting the impacts of a policy on many groups of the population, and aggregating the results. Alternatively, the same models can be used to obtain a crude prediction of the impacts of the same policy. The analyst or planner will make the appropriate selection of procedures, considering the required level of accuracy (or sensitivity to errors), cost, turn-around time, etc.

The suggested methods also vary in the level of spatial detail to which they can be applied. The impacts of changes in operating policies on a single bus route, such as scheduling, fare structure or others, are at the lower end of spatial detail. Metropolitan policies, such as pricing or priority treatment of high-occupancy vehicles, may lie at the upper end of spatial detail.

Two other dimensions are demonstrated by the examples given above. These are the time horizon and the type of transportation options analysed. The behavioural modelling approach has been successfully applied in the analysis and prediction of impacts of projects involving construction of facilities or new modes of transportation, as well as projects involving operating changes or traffic management changes. Obviously, these differ in the time horizon of their impacts as well as in type of impacts.

At any level of analysis the models used are basically the same, but aggregation and the specific procedures of data preparation will vary.

Modelling options. Assuming that the theory of choice underlies travel behaviour, one can easily observe that a trip is a result of a choice process along several dimensions. The choice of travel mode, for example, is not independent of other choices made simultaneously by the decision-making unit (an individual or a household). These may include long-term choices, such as residential location and automobile ownership, and short-term choices of destination, the frequency of trip making or the time in which to make the trip. Obviously, the interdependence of these choices is different for various trip purposes. For most people the daily trip to work does not involve frequent changes of destination, frequency or time of day, but trips for shopping or recreation do involve more frequent changes of this whole set of choices. In any given context, the structure of the choices made needs to be explicitly identified and different choice sets, relevant to the issue at hand, will be included in the model estimation. The initial development of disaggregate models treated only a binary choice structure, specifically the choice between two modes of travel. Since then, a series of models, each specifically estimated for a relevant set of choices, has been developed.[11] [12] [13]

One type of input, common to all models, is the level of service data. By their nature, these relate to the performance of a specific route or network. Yet, in many cases, collecting and managing a data set on a complete network may exceed the budgetary constraints for the whole analysis. Two basic approaches to treat the level of service data have been employed. In the first, where network data are maintained or the analysis is limited to a corridor or a single route, the actual network or link is used. Examples of such cases are metropolitan areas which keep a data set for a complete Urban Transportation Planning System (UTPS) and the disaggregate choice models are incorporated into this system; or when an analysis of a preferential treatment project on a single artery is performed, the actual corridor network may be modelled. The second approach is to represent the network in some abstract form, i.e. using a less detailed, simplified representation of some links and nodes of the network. There may be different levels of abstraction, depending on the level of detail required in the output. Any abstract representation of the network entails a set of assumptions which constrain the validity of the results, however. These assumptions need to be made explicitly so that accuracy of conclusions can be evaluated.

Another modelling option open to the analyst is the inclusion or exclusion of an explicit equilibration procedure. Theoretically, an equilibration of the demand and level of service supplied is an essential step, but it is usually complex and expensive. Practically, the analyst may decide to omit this step when the level of accuracy of the results does not warrant it or when he has reason to assume that the level of service used as input approximates the equilibrium conditions, i.e. there are no serious congestion effects.

Simplified models and sketch planning. Sketch-planning procedures are designed to perform a quick examination of a large number of alternative policies. Such analysis tools have been developed over the past few years for various planning purposes, ranging from the study of national urban transportation resource allocation[14] to the preliminary screening of alternative transportation system configurations at a sub-area level.[15] Urban transportation sketch planning packages are generally characterised by a high degree of geographic aggregation and network abstraction, limited information requirements, ease of input data preparation and fast response times. Reviews of recent research in sketch-planning methodologies are given by Landau[16] and Watanatada.[17]

Recently, some effort has been directed to the development of simplified analysis methods, which would be compatible with the needs or constraints of sketch planning but utilise the advantages of disaggregate travel demand models. Specifically, two properties are vital in this approach. The models need to maintain sensitivity to the policies considered in today's political, social and economic environment. Secondly, the most promising approach is to reduce data collection and management costs, in view of budgetary constraints. Simplified analysis methods can be computer-based or even performed by programmable pocket calculators or manual pencil and paper methods. The following section will demonstrate two simplified analysis methods developed in the past three years.

One of the difficulties in applying disaggregate travel demand models to sketch planning is the problem of developing concomitant transportation supply and traffic assignment models for various levels of geographic aggregation. One fundamental modelling difficulty is that because a transportation network is not just a direct summation of the individual links, the problem of network aggregation becomes analytically intractable. Because there is no consistent theory for network-abstract supply modelling, researchers have developed relationships based on experimental results or highly simplified assumptions on transportation supply characteristics. For a given zone pair, the supply models should predict the distribution of level-of-service attributes as a function of travel demand and network capacity.

Because of the extremely wide range of geographic aggregation employed in sketch planning, transportation supply and traffic assignment models should be developed in integrated network and network-abstract forms. Network supply models represent transportation facilities, mostly of major types, as network links. Network-abstract supply models represent transportation facilities, mostly of ubiquitous nature, as aggregate transportation systems defined by mode, facility type and geographic unit. A network-abstract model can be developed to relate parametrically the aggregate performance measures of a system to its transportation supply characteristics and traffic loads. Network-abstract supply models can be developed for both access and line-haul travel. Talvitie and others have developed network-abstract supply models for the access portion of a trip.[18] [19]

Network traffic assignment models for sketch planning can be developed largely on existing knowledge. Network-abstract traffic assignment models can be developed to allocate traffic, not to routes, but to aggregrate transportation systems.[20]

EXAMPLES OF SIMPLIFIED ANALYSIS METHODS

Within this class of methods one would also find a wide range of possible approaches to the analysis of urban transportation problems. What all these methods share is the sensitivity to the policy options with which the planner is confronted, and the continuous effort to improve the efficiency of data utilisation.

Two methods developed within this class will be briefly described in this section. The first is a computer-based model labelled MIT-TRANS,[21] [22] and the second is a method based on programmable pocket calculators in the Responsive Analysis Methods Project (RAMP) at MIT.[23] [24]

The MIT-TRANS model. This computer-based model exemplifies one approach in using disaggregate models within the strict limitations of sketch-planning data requirements. A two-pronged approach is suggested: first, to represent the distributions of the independent variables as parametric functions of readily obtainable aggregate data; and secondly, to employ Monte Carlo simulation methods in forecasting aggregate travel demand. Monte Carlo methods are employed in two stages of sampling. The first stage generates a sample of households distributed over the urban area. The second stage samples a set of potential destinations by trip purpose for each household in the sample. Travel forecasts for each household are then computed on the basis of these potential destinations.

The Monte Carlo simulation approach has three important advantages. First, it is not restricted to any type of mathematical presentation. Secondly, its prediction errors are not subject to aggregation bias and can be easily controlled. And thirdly, computational experience has shown it to be relatively inexpensive to produce the kind of forecasting precision required for many urban transportation planning purposes.

The model treats the whole urban area as one unit, yet the predictions can be disaggregated to some market segments to allow the analysis of the incidence of the impacts. The general structure of the model is shown in Figure 1. While this model treats only the demand side, it still allows the analysis of a wide range of policy options, such as pricing for transit or car users and car-pool promotion. The model treats a number of trips and includes seven separate disaggregate models which are linked together, as shown in Figure 2.

The output of MIT-TRANS includes the number of trips made, the mode shares, person-miles and vehicle miles travelled, average vehicle occupancy by trip purpose, and various other aggregate predictions. Running MIT-TRANS on a computer is relatively cheap. Test runs for the Washington DC area required less than one minute of CPU (on IBM 370/168) with standard errors of less than two per cent of the mean forecasts.

RAMP procedures. Another simplified approach is to use programmable pocket calculators (PPC). These machines provide a complete computer logic but lack the ability of treating large data sets. Programs as well as data can be recorded on magnetic cards which are fed into the PPC. The calculators are cheap (the most advanced model sells for less than $230) and easy to operate. A library of PPC programmes has been developed for a variety of transportation analysis needs. It includes programmes for prediction of demand (employing disaggregate choice models), operating impacts, performance and environmental impacts, and a number of utility programmes. (A complete list of currently available programs is given in Appendix A.)

RAMP procedures can be applied in different planning contexts, ranging from an area-wide quick and crude prediction of the impacts of transportation policies, to a detailed analysis of a single project. The procedures are appropriate for the analysis of most Transportation Systems Management (TSM) because they are inexpensive, require modest data inputs, are simple to perform and provide fast responses.

A case study application demonstrating RAMP procedures has been developed to predict the demand for car-pools (of three or more occupants) and buses following the introduction of an exclusive lane for these modes on Boston's south-east expressway.[24] A disaggregate mode choice model estimated elsewhere was transferred and adjusted to the Boston case.[25] Tract level from the US Census was the basis for the socio-economic data needed in the disaggregate mode choice model. Level-of-service data were obtained from transportation agencies in the region. No field data were collected. The data (both socio-economic and level-of-service) for each of the 140 tracts from which travellers use the south-east expressway corridor were recorded on magnetic cards.

The analysis procedure is illustrated in Figure 3. Stage B is the development of the Base Case, i.e. the prediction of the pre-project situation, which is necessary for the adjustment of the transferred mode choice model. Implementation of the policy (Stage C) is represented in the demand model by reducing travel times and costs for the preferred modes (car-pools and buses) and increasing travel time for the non-preferred modes, which are expected to experience higher congestion effects owing to the reduction in the number of lanes on the expressway. The policy is likely to have a number of impacts, apart from the change in mode choice. Car-pools which have used alternative routes may be diverted to the preferred facility, non-preferred modes may divert to alternative routes. Trip makers may change the time of their trip or cancel their trips. Each of these impacts may have a different magnitude and the analyst should make a decision on which should be analysed and which may have only marginal importance. In this case study (Stage E) only route diversion was analysed. Modal shares under the policy were predicted within about 3 per cent of the observed behaviour. The procedure, including data preparation but excluding programming time, requires two to three work-days.

REFERENCES

[1] Binder, R.H., Major issues in travel demand forecasting in "Brand and Manheim" (eds) (1973).

[2] Bouchard, R.J., Relevance of planning techniques to decision-making in "Brand and Manheim" (eds) (1973).

[3] McFadden, F., "Conditional logit analysis qualitative choice behaviour" in Zarembka, Paul (ed) Frontiers in econometrics (1974)

[4] Manski, C.F., The analysis of qualitative choice (1973), unpublished PhD dissertation, Department of Economics, MIT.

[5] Atherton, T.J., and Ben-Akiva, M., Transferability and updating of disaggregate travel demand models (1976), TRR 610, TRB.

[6] Koppelman, F.S., Guidelines for aggregate travel prediction using disaggregate choice models (1976), TRR 610, TRB.

[7] Cambridge Systematics, Inc., Car-pooling incentives: analysis of transportation and energy impacts (1976), prepared for Federal Energy Administrators, Washington, DC.

[8] Voorhees, A.M., and Associates, Inc., Cambridge Systematics, Inc., and More-Heder Associates, Auto restricted zones (1977), Final report, UMTA

[9] Barton-Aschman Associates, and Cambridge Systematics, Inc., Los Angeles central business district: internal travel demand modelling (1976), prepared for Community Redevelopment Agency, Los Angeles, California.

[10] Spear, B.D., Applications of new travel demand forecasting techniques to transportation planning - a study of individual choice models (1977), Federal Highway Administration, US DOT.

[11] Adler, T.J., Modelling non-work travel patterns (1975), PhD dissertation, Department of Civil Engineering, MIT.

[12] Adler, T.J., and Ben-Akiva, M., A joint frequency, destination and mode model for shopping trips (1975), TRR 569, TRB.

[13] Ben-Akiva, M., "Passenger travel demand forecasting: applications of disaggregate models and directions for research" in Transport decisions in an age of uncertainty (1977), Boston.

[14] Weiner, E., Kassoff, H., and Gendell, D., Multimodal national urban transportation policy planning model (1973), HRR. 458

[15] Dial, R.B., A procedure for long range transportation (sketch) planning (1973), Proceedings of the International Conference on Transportation Research, Transportation Research Forum, Bruges, Belgium.

[16] Landau, U., Sketch planning models in transportation systems (1976), unpublished thesis, Department of Civil Engineering, MIT.

[17] Watanatada, T., Application of disaggregate choice models to urban transportation sketch planning (1977), unpublished PhD thesis, Department of Civil Engineering, MIT.

[18] Talvitie, A., and Hilsen, N., "An aggregate access supply model" (1974), Trans. Research Forum Proceedings, Vol 15, No 1.

[19] Talvitie, A., and Leung, T., Parametric access network model (1976), TRR 592.

[20] Creighton, Hamburg Inc., Freeway-surface arterial splitter (1971), prepared for FHWA, US DOT.

[21] Watanatada, T., and Ben-Akiva, M., Spatial aggregation of disaggregate choice models: an area-wide urban travel demand sketch planning model (1978), presented to the TRB.

[22] Watanatada, T., and Ben-Akiva, M., Development of an aggregate model of urbanised area travel behaviour: final report, US DOT (1979), Assistant Secretary for Policy and International Affairs and FHWA.

[23] Manheim, M.L., Furth, P., and Saloman, I., Responsive analysis methods: transportation analysis using pocket calculators (1978), Working paper CTS - RAMP - 77-1, MIT; also, TRR in press.

[24] Salomon, I., Application of simplified analysis methods: a case study of Boston's South-east Expressway car-pool and bus lane (1978), unpublished MS thesis, Department of Civil Engineering, MIT.

[25] Cambridge Systematics, Inc, MTC Travel Model Development Project: Final Report Volume I: Summary Report; Volume II: Detailed Model Descriptions; Volume III: MTCFCAST Users' Guide (1977), Report for the Metropolitan Transportation Commission, Berkeley, California.

OTHER READING

[a] Ben-Akiva, M., and Atherton, T., Methodology for short range travel demand predictions: analysis of carpooling incentives (1977), Journal of Transport Economics and Policy, 11:3, pages 224-61.

[b] Difiglo, C., a d Reed, Jnr., R.F., Transit sketch planning procedures (1975), TRR 569, Washington DC, TRB.

[c] Dunbar, F.C., Quick policy evaluation with behavioural demand models (1976), TRR 610, Washington DC, TRB.

[d] Liou, P.S., Cohen, G.S., and Hartgen, D.T., An application of disaggregate mode choice models to travel demand forecasting for urban transit systems (1975), TRR 534.

[e] Small, K.A., Bus priority, differential pricing, and investment in urban highways (1976), WP 7613, UTDFP, Institute of Transportation Studies, University of California.

[f] Train, K., Work trip mode split models: an empirical exploration of estimate sensitivity to model and data specification (1976), WP 6602, UTDFP, Inst. of Transp. Studies, University of California.

Abbreviations used in references are: DOT (Department of Transportation), FHWA (Federal Highway Administration), HRR (Highway Research Record), MIT (Massachusetts Institute of Technology), TRB (Transportation Research Board), TRR (Transportation Research Record), UMTA (Urban Mass Transit Administration), and UTDFP (Urban Transportation Demand Forecasting Project).

APPENDIX: PROGRAMS FOR PROGRAMMABLE POCKET CALCULATORS DEVELOPED
BY RAMP

Program title	Acronym	Calculator(s)
Prediction - demand-mode choice		
1. Incremental changes in model demand-pivot-point with multinomial logit:	[PPMNL-1(A)]	HP 67, 97
2. The same	[PPMNL-2(A),(B)]	TI 52, 59
3. Binary mode choice for work trips	[2 MODE-2(A),(B)]	TI 52, 59
4. Binary mode choice (logit) with aggregation work trips:	[2 MODE-AGG-1(A),(B)]	TI 52, 59
5. The same	[2 MODE-AGG-2(A)]	HP 67, 97
6. Three-mode choice with aggregation work-trips	[3 MODE AGG-1(A)]	TI 52
Prediction - operator impacts		
7. Operator impacts - pivot-point with multinomial logit	[OPERATOR-1(A),(B)]	HP 67, 97
8. Operator impacts - parametric analysis with multinomial logit	[OPERATOR-2(A),(B)]	TI 59
9. The same	[OPERATOR-3(A)]	HP 67, 97
10. Operator impacts - parametric analysis with plotting by multi-nomial logit	[OPERATOR-4(A)]	HP 67, 97
Prediction - performance		
11. Vehicular travel time	[TT-1(A),(B)]	TI 52, 59
12. Intersection delay	[INTERSECTION-1(A),(B)]	TI 52, 59
13. Intersection delay	[INTERSECTION-2(A)]	HP 67, 97
14. Bus ride simulation	[BUS-2(A)]	TI 52, 59
Prediction - equilibration		
15. Congested equilibrium on a single link	[EQ-1(A)]	HP 67, 97
Prediction - environmental effects		
16. Bus air-pollutant emissions	[BUS POL-1(A),(B)]	TI 52, 59
17. Energy consumption on a bus route	[ENERGY-1(A)]	TI 52
Inferences about context		
18. Origin-destination matrix from on/off counts: Version A	[ODA-1(A)]	TI 59
19. Origin-destination matrix from on/off counts: Version B	[ODB-1(A)]	TI 59
20. Generation of sample households	[HHGEN-1(A)]	TI 59
Utility programmes		
21. Matrix updating with iterative proportional fit	[IPF-1(A)]	TI 59

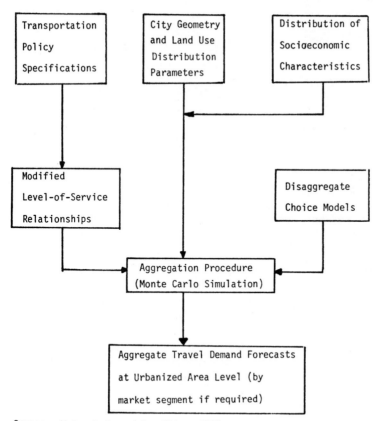

Source: Watanatada and Ben-Akiva, 1979.

Figure 1. Basic operations of MIT-Trans model

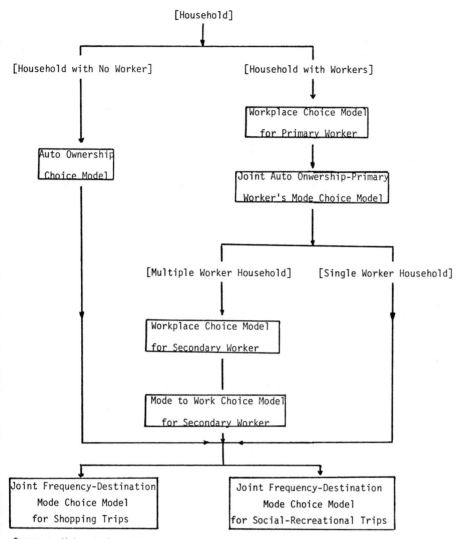

Source: Watanatada and Ben-Akiva, 1979.

Figure 2. MIT-Trans disaggregate choice models

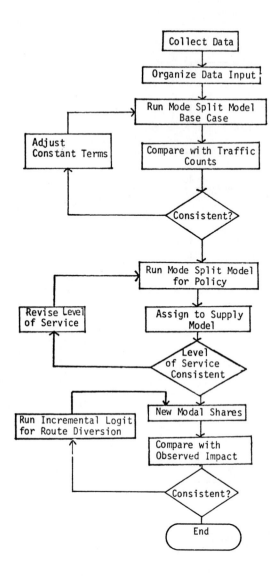

Figure 3. Analysis structure

The functional approach to transit balance

11

Eric L. Bers   United States

SYNOPSIS

For planners, operators and politicians, there is no simplistic, accurate and reliable decision-making tool for selecting a service mode for a specific region, corridor, area or route. The primary objective of this paper is to define transit systems on the basis of the type of service offered and its relationship to the accompanying (supporting) transit modes. An underlying purpose is to develop a standard or guideline to assist the providers of public transport in establishing an economical, convenient and balanced system of transit services.

This paper presents a comprehensive approach to transit planning and operations based on corridor travel requirements and the three basic trip components: collection, line-haul and distribution. Each segment of the trip must be so designed that it will permit a coordinated modal interface and provide service that can compete with other forms of transport. The functional approach to transit balance requires that each transit service (route) be defined in terms of its function: movement versus access. As in the highway mode, a specific transit route canot offer rapid movement concurrently with land access. The substance of this paper is founded on the operational definition of a transit mode; each modal alternative is explicated in terms of the degree of right-of-way control, the type of operation and tye type of technology. The application of this procedure provides a unique (precise) description of transit services.

The evolution and development of a transit system in a corridor is discussed. When the corridor travel needs are determined, the supporting ridership area must complement the corridor service levels. Thus, service function is matched with demand. The product of this approach is a series of operation guidelines for transit system balance. Each level of movement and access is defined for various right-of-way configurations and categories of operating mechanisms. Service functions and their accompanying technologies are indicated. The main thrust is that any particular type of vehicle has more than one service function under different operating schemes or right-of-way classifications.

INTRODUCTION

The primary objective of this paper is to define transit systems on the basis of the type of service offered and its relationship to the accompanying (supporting) transit modes. An underlying purpose is to develop a standard or guideline to assist the providers of public transport in establishing an economical, convenient and balanced system of transit services.

Alternative analyses or mode comparisons can be reoriented to include inherent issues associated with transit system balance and transit service quality. More efficient systems can be operated by productively integrating the different transit sub-systems. This approach is applicable to corridor studies as well as regional analyses. Since regional transit systems are composed of many distinct transit corridors with varying degrees of modes (or services), this paper suggests that regional systems be analysed at the corridor level. When they are combined, the transit corridor or sub-systems add up to the full regional system. Some adjustment may be necessary in balancing parallel and competing facilities.

The information flow of this paper traces transit balance in relation to three determining factors: movement versus access, definition of a transit mode, and evolution of a corridor. Prior to the discussion of these factors, background data are supplied. The functional trade-offs between movement and access are developed on the basis of travel speed, passenger flow rate, reliability and the frequency of flow interruptions. Definition of a transit mode explains a modal alternative in terms of the type of right-of-way (ROW) control, the type of operation and the type of technology. The evolution of a corridor describes the relationship through time between transit system development and trip making. Subsequently, a series of guidelines for transit balance is presented.

Perspective. Alternative analyses are undertaken to determine the best travel mode for a site-specific corridor. Unfortunately, selection of modes is, more often than not, based on technological considerations, i.e. bus versus rail. This problem is further compounded by the fact that the Urban Mass Transit Administration (UMTA) of the US Department of Transportation offers large quantities of money for specific technology demonstration projects such as Downtown People Movers (DPM) and Light Rail Vehicle (LRV) systems. Cities seeking 20-cent dollars (federal funds are provided on an 80/20 matching formula) are encouraged to compete for and install new projects which may not solve current or projected urban transport problems. Since UMTA has a thorough review procedure for all grant proposals, the misuse or inappropriate application of technology is greatly reduced. Instead of arguing about bus versus rail in mode comparisons, transport planners should initially investigate the type of service (function) needed in a corridor (or an area) and then select the system best suited in terms of cost and other factors to provide that service.

According to UMTA, a regional multi-modal "comprehensive strategy may involve the construction of a rail rapid transit line in a corridor of heavy demand, supplemented by a light rail network or bus-ways in a lower density portion of the metropolitan area, and assisted by fleets of flexibly routed para-transit vehicles acting as suburban feeders to the high capacity line-haul systems".

The approach outlined above follows the proper path of transit balance between service functions, but the statement implies many facets of transit balance without ever directly indicating so. In this regard, the statement is masked. It is important to identify and fully understand these functional relationships. In the context in which it was intended, this statement is a fine example of sound logic as it relates to balanced transit systems.

A balanced transit system is directly and uniformly applicable to private automobile travel. Alternative analyses for highway projects are not currently performed as rigorously as required by UMTA, but project comparisons are made between various roadway facilities. For example, the trade-offs between a freeway and a major arterial are typically evaluated. The roadway system for automobiles is balanced with freeways, expressways, major and minor arterials, collectors, local streets and culs-de-sac. As the urban area expands, the travel needs of the corridor dictate the type and extent of facility. This process has been implicitly followed throughout the years.

Applied to any system, the three components of a trip are collection, line-haul and distribution. Trip collection is the initial process of accumulating or collecting the passengers on to the vehicle. This segment requires a high degree of access to contiguous properties. The line-haul portion of the trip consists of a large passenger movement toward the destination with few if any stops. Travel speed is the highest on this leg of the trip. The distribution segment systematically delivers the riders to their respective destinations; it is the negative or complement of collection.

Travel requirements are unique for each portion of the trip and different types of transit services are needed to fulfil the needs of each stage of the trip which must be individually and specifically designed to permit a coordinated modal interface.

Each type of service will perform a different function to complete the system. The broad categories of service types generally consist of local, circulator, express and regional routes. Local routes are the backbone of any system and provide the bulk of service. Local routes are subdivided into two distisnct groups, radial and circumferential. Radial local routes originate, terminate or pass through downtown areas while cross-town or circumferential local routes do not.

Circulator routes within neighbourhoods provide connections to high activity centres, such as shopping centres, schools, and employment centres. A second function of the circulator is to feed local and express services. Express and regional routes are high-speed trips which provide transportation from transit centres and neighbourhoods in outlying portions of metropolitan areas to major employment centres, primarily downtown. Expresses run basically during morning and evening peak hours. These transit services are not functionally different from those in the highway mode, but the fundamental commonality is so understated. The section on movement and access begins to explain these relationships.

MOVEMENT VERSUS ACCESS

The functional approach to transit balance requires that each transit service (route) be defined in terms of its function: movement or access. Traffic service and land access are necessary but conflicting functions of any transportation system. As in the highway mode, a specific transit route cannot offer rapid movement as well as efficient access. Satisfying both functions adequately requires a variety of transit services or modes. At one extreme, a heavy rail transit route with lengthy inter-station spacings limits access points and most effectively provides for high design speed. At the other extreme, a bus operating locally in mixed traffic provides maximum access to abutting properties at the expense of speed or free-flowing movement. An analogous situation exists for road-based automotive traffic where a freeway and a cul-de-sac or local road exhibit the same characteristics respectively.

Transit balance can only be achieved when each service functions in its own domain. The movement function provides an opportunity for user mobility in predominantly high travel speeds and a large passenger flow rate (passengers/hour). Consequently, route capacity and reliability are high and travel time is short. The access function offers ease of ingress/egress (entry/exit) for local properties. The level of access control regulates the magnitude of travel opportunities available to the transit user. Travel speed is typically low and hence reliability and capacity are low; travel times are generally long.

Figure 1 demonstrates the division of these functions for various functional classes of transit. The curve is labelled frequency of traffic interruptions to delineate the trade-offs possible in choosing between movement and access. The similarity of this relationship to that of automotive travel is not coincidental. In fact, the transferability is appropriate and exact. The various types of roadways, such as local streets, collectors, arterials and freeways could be placed on the graph in that order to convert this relationship to that based upon the automobile. The relationship is inherently identical for automobiles as transportation operations are currently practised. The reason is simple: as explained earlier, direct access to abutting land is contradictory to rapid movement. This matter is addressed later in the paper.

The extended meaning of frequency of traffic interruptions includes transit right-of-way and type of operation. A transit vehicle running over exclusive (private) right-of-way experiences limited traffic interruptions. Similarly, a vehicle operating express over a portion of its route functions in an identical manner. Thus, access and movement can be regulated by both the extent of right-of-way protection and the type of operation. This concept is crucial in determining and selecting the proper functional balances for a transit service.

From the viewpoint of both transit operator and user, a mode change or transfer can be an effective tool in improving transit service. Typically, a feeder bus service is employed to join a rail station, thus causing a mode change. A variance of this service might include a feeder bus service which collects and distributes riders and then runs express to its destination on a bus-way. The entire trip is now completed within the same vehicle. In other words, collection, line-haul and distribution can be accomplished on a single vehicle. The collection and distribution segments of the trip are predominantly access functions, while the line-haul portion is primarily a movement function.

A transfer is introduced at the functional interface (juncture) of the collection/distribution and line-haul portions of a trip or at line-haul to line-haul extensions. The intersection of two or more travel modes provides a strong potential for passenger interchange (transfer) and expands the travel opportunities (area of coverage) for all its users. Trip making may be induced by these greater opportunities. The key element or benefit of transfers relates to their location. If the transfer capability is at a high activity centre, e.g. shopping centre, more trips are generated.

Transfers between modes can be executed in all cases provided there is proper planning and coordinated design. To complete all trips, a suitable combination of movement and access is required. During each portion of the trip, the specific requirements of the user change dramatically; thus the interconnection between facilities is essential to join these trip ends and complete the total trip. The transfer can be employed to link the line-haul modes (movement) with the collector/distributor modes (access).

For short trips, a mode employing predominantly access functions is acceptable and desired. Long trips, however, need primarily the movement function to reach their destination. A transfer may occur while completing any one trip, but current technology is sufficient to permit an entire trip to be performed in a single vehicle. The bus, trolley-bus and light rail vehicle (LRV) are among the transit modes most suited to the completion of a trip without transferential between collection, line-haul and distribution. Thus, fixed and non-fixed transit modes can provide the proper balance between movement and access. In certain high-density areas, such as New York City, transfers are of the rail-to-rail type which is performed rather conveniently. For lower volume trip interchanges, transferential is essential to reach many destinations. A mode change should be viewed as a positive means to join low-volume interchanges. Although the transfer is an impediment to certain rider groups, it can be utilised to one's advantage if it is coordinated properly. Modes which can be employed to transport riders in the same vehicle should be re-examined and their services expanded to exercise this positive trait.

DEFINITION OF A TRANSIT MODE

In the operational sense, there are three primary elements of a transit mode: right-of-way category, operating mechanism, and technology consideration. Classifying transit modes solely by technology, as is often done, is incomplete and frequently leads to ambiguity. After all, a light rail vehicle (LRV) and a bus differ greatly in technology, yet they may both operate over the same right-of-way category and with the same operating mechanism. Hence, the typical user might note the obvious technology/vehicle differences, but would never sense the distinction between the transit services. (This matter will be more adequately explored in the section on operational guidelines.) For this reason, it is necessary to define each transit mode comprehensively to eliminate misinterpretation and minimise inappropriate applications.

The types of transit system rights-of-way can be classified into three mutually exclusive categories: systems operating on surface streets in mixed traffic, primarily offering access to adjacent land; systems operating on partially controlled right-of-way, providing a balance between movement and access; and systems operating on fully controlled or grade-separated right-of-way predominantly for movement.

Each successive category actually represents an increase or step in transit service quality. Transit vehicles running in mixed or local traffic are subjected to the same obstacles and delays (inclement weather, accidents, finite and limited roadway capacities) that plague the automobile. The second category offers an improved degree of service quality through capacities for higher speed, reliability and capacity and has more control over traffic interruptions. As a result travel time by transit can be competitive with that of the private automobile. The highest quality of service in all respects is provided by vehicles operating in the third category where the entire length of the right-of-way is grade-separated. The high speed and large passenger flow rate in this category offer the transit user a faster higher-capacity trip than does a car from point to point.

The three classifications of right-of-way can be illustrated with road-based modes. In the first category, vehicles would operate on local streets. In the second category, transit vehicles would operate in transit lanes in mixed traffic and, at a certain intersection, would be given preferential treatment (i.e. grade-separation, non-restricted turning movements, or green signal preemption). Vehicles running in the third category would operate on exclusive private rights-of-way (i.e. busway).

The operating mechanism of a mode describes the stopping frequency of inter-station spacing that the transit vehicle encounters. The three types of vehicle operating mechanisms are as follows: local running or stops every block; limited-stop or skip-stop where a portion of the trim is a closed-door operation; and express service where nearly 100 per cent of the trip is line-haul and there are very few stops. In this order, each subsequent type of operation represents an increase in transit level of service. Local transit vehicles are characterised by frequent stops, low travel speeds and low route capacities. Limited-stop vehicles have higher travel speeds and passenger flows, while express operations provide the highest travel speeds and greatest passenger flow rates. As in the previous discussion on right-of-way categories, the first category (local stops) is used primarily for access to abutting land, the second category (limited-stop) provides a balance between land access and passenger movement, and the third category (express service) offers predominantly movement. Theoretically, any particular operating mechanism can be employed on any type of right-of-way. Thus, provision of a transit service is open to a wide selection of operating conditions over many different types of right-of-way.

The final variable in the definition of a transit mode is the technology. Technology considerations include such matters as vehicle types, rail or rubber-tyred guide-ways, highway-based modes, and automatic train control (ATC). These parameters define seating/standing capacities, acceleration/deceleration rates, operating costs, and minimum permissible headways. The selection of technology is explored further in the last section.

It is also necessary to indicate for the entire length of the trip the service variations during collection, line-haul and distribution. The use of this logic in discussing transit modes clearly generates a unique (precise) description of service. A comprehensive definition might be as follows: a rubber-tyred bus operating locally in mixed traffic during collection, and then running express during the line-haul segment of the trip with exclusive guide-way controls (busway), and finally operating with limited stops in mixed traffic for down-town distribution.

EVOLUTION OF A TRANSIT CORRIDOR

The evolution of a transit corridor follows a typical series of events based on the accompanying land use and road network. As rural or outlying urban areas begin to change (grow), a definite road network begins to take shape. Local and collector facilities expand into arterial roadways and arterials are upgraded to freeway status. Freeways are occasionally constructed parallel to other arterials. The major objective of this road-building exercise is to link the growing area to a city or high activity centre with a higher movement facility.

These changes can be expected to occur as a settlement grows into a small town, then to a city and finally into a large urban area. Once a natural corridor is established, the magnitude of trip interchanges continues to expand and the land use changes to higher densities. As land densities intensify, travel requirements increase. Transportation facilities expand to match the needs created by the development activities. Traffic signals replace stop signs at selected intersections and grade-separated interchanges are constructed at the points of highest volume.

As the volume of the trips increases, the transit system develops accordingly. Initially trip making is dispersed, but certain specific activity centres attract trips, providing the first opportunity for group travel. Car and van pools spring up spontaneously in the beginning, followed by a subscription bus operation. As more trip making occurs, conventional transit commences operation. Preferential treatment such as transit lanes may be employed to speed transit vehicles, particularly in downtown areas. With more growth, some of the right-of-way becomes grade-separated (fixed guide-way), using buses or light rail vehicles. Express and limited-stop vehicle operations are instituted.

Trip lengths are long (downtown) and the system is a prime candidate for fully controlled right-of-way. A higher level of transit service is provided within this type of service. More varied types of services are needed to cope with expanding trip making. Automatic operation of vehicles on fixed guide-ways is feasible. Thus, a transit system begins with a very low capacity service and expands as travel requirements change. Service quality and system capacity increase with larger passenger volumes. This sequence is best described as transit system development.

In a radial or circumferencial corridor, the resulting adjacent land use may be residential or industrial/commercial, and thus producing or attracting trips respectively. To attract patrons along the trunk line, irrespective of its type of service (mode, right-of-way, operation), the adjacent land must be serviced by a local (collector or distributor) mode of transport. If walking distances are within 1,200 feet, the pedestrian mode is preferred. In an area with substantial density and excessive walk distances to a transit facility, a feeder service is sensible. These sub-systems (circulator) may include surface or trolley-buses or light rail vehicles, depending upon demand volumes. Automobile access remains the primary mode of access, with parking in nearby residential neighbourhoods. Informal parking at shopping centres is also widely employed. When the weather is inclement, walking distances are long and little or no transit service is available, there is generally an excess of automobile usage.

A grid-like street system provides the most service opportunities for transit and pedestrian alike. With a series of parallel and perpendicular streets, the transit operator can offer services (routes) at two or three-block intervals. Walk access is minimised and the user is typically able to reach any particular destination with at most one transfer. This technique is employed in downtown areas where the street system affords such treatment of service. The outlying areas or fringes of downtown contain a street system entirely different in structure, however. Primarily for that reason, auto and feeder bus access to transit is more prevalent than pedestrian access.

Between two or more parallel or distant corridors, another corridor may develop to join each one at some activity centre, e.g. shopping centre. As the open and available land is filled in and density builds up, exclusive movement between corridors is no longer necessary and a balance between movement and access for the intermediate land use is more appropriate. At this juncture, feeder services are introduced to the accompanying or supporting ridership areas. These services increase use of the line while reinforcing the adjacent development.

Fixed facilities such as a LRV or busway system can shape development patterns. Transit does not always have to follow development and can be employed as a tool to structure future land use through fixed guide-way facilities or other transit services. The present state of urban transit systems is such, however, that very little operating and capital money is available to induce development in a new area. Certainly, if transit services were to be extended into an undeveloped, but potential area, land use changes would occur. This concept does not differ from the construction of a highway facility into a vacant corridor (undeveloped area). After several years, the residents seem unable to explain the population explosion and land use effects. If accommodated in the proper fashion, a transit corridor could evolve in the same manner as a highway corridor.

OPERATIONAL GUIDELINES

The compilation of information in the preceding sections has provided a firm understanding of the functions of transit service relationships and objectives. This section synthesises operational guidelines. Table 1 summarises a qualitative assessment and comparison of each distinct transit function. Each facility function refers to the right-of-way control and operating mechanism, since both access and movement depend on these variables. As noted previously, reliability, capacity and speed increase with fewer traffic interruptions. Collector/distributor passenger modes (access function) typically entail short trips over short station spacings. Line-haul travel is employed to link the CBD, major generators and secondary generators with distant production bases; collection/distribution travel joins individual sites and local areas to transit services.

Table 2 presents operational guidelines for use in developing transit systems. The local and limited-stop categories are employed as collector/distributor modes and the express and limited-stop operations are used to provide line-haul services. Similarly, partially controlled and uncontrolled right-of-way is a feature of collection and distribution, whereas fully controlled and partially controlled ROW represent line-haul operations. Local operations and uncontrolled rights-of-way are predominantly used for access. Express mechanisms and fully-controlled ROW are considered primarily for movement. Between these services are limited-stop operations and partially controlled rights-of-way; they comprise a mixture of movement and access and are useful in either extreme.

Alternative analyses should be conducted to determine the type of service functions needed before the application of technology. The selection of technology should be founded on the requirements of rights-of-way and operating mechanisms. Service comparisons need to be made at different levels of transit service, and then a detailed analysis is appropriate within a particular ROW/operating mechanism category. It should be apparent now that a rubber-tyred bus operating locally on a city street for collection, then on a grade-separated busway express in the line-haul mode, and finally locally in mixed traffic during down-town distribution is functionally no different from a LRV running locally over a city street during collection, then express on a grade-separated ROW in line-haul, and finally again locally in mixed traffic in downtown distribution. In terms of the service function, the final modal selection is actually based on inherent technological and operating advantages such as travel speed (time), operating cost, capital cost, system capacity, reliability and passenger attraction.

The success of this procedure is reflected in the generality, and hence adaptability of the results to other systems. Operational problems are reduced to the level of service to be supplied, and no longer to a conflict between bus and rail or light van and heavy rail services. Once the service level is defined, local decision makers can focus on the operational trade-offs. The final technological decision is virtually immaterial, since the proper service level has been matched with the demand.

Table 2 is productive in the evolution and development of transit systems. As travel demands increase and the area of a corridor expands, more diversified and specialised services (modes) are needed. Some sub-systems are converted to higher capacity modes while new services are implemented. An LRV system is briefly discussed to illustrate this point. Initially, a streetcar operation may exist with local stops each block. When trip distances lengthen and passenger volumes rise, limited-stop or express operations may be introduced to accommodate the various types of users. Another alternative might include partially controlled ROW which may eventually be upgraded to fully controlled ROW. Thus, there is never inflexibility in the selection of one technology over another. Although the technology remains constant, the variance in right-of-way categories and operating mechanisms is so great that the previous service may not resemble the present service mode, i.e. streetcar and grade-separated LRV.

Another product of these charts is that fixed and non-fixed guide-way systems have specific domains of operation. The trade-offs between movement and access are nearly identical to those regarding the type of guide-way. The highest degree of movement is achieved with fixed guide-way systems while the greatest level of access is realised with non-fixed systems. In very dense cities, such as New York, this approach needs some slight modification. Fixed guide-way systems operate locally, skip-stop, and express. The skip-stop operations provide a suitable degree of access to nearby properties and the interface with local rapid rail transit completes the travel needs of the majority of users.

For transit to be competitive with the automobile, the ROW category and operating mechanism are critical elements. As indicated in the section on the definition of a transit mode, private rights-of-way and express operations from point to point can be effectively employed to ensure that transit is superior (in terms of time) to the private automobile. A transit corridor can transport more passengers by a given location than any type of highway facility for a specific right-of-way width. Automatic train control (ATC) can be used to increase safety, capacity, speed, and reliability. With all these benefits, what are the disadvantages of transit? A large part of the problem is trip distribution from the line-haul mode. While an entire trip may be accommodated without transferring, not all destinations are readily available from any specific route. A transfer to an extensive circulator (distributor) mode can generally solve the problem. The Downtown People Mover (DPM) is such a mode. It travels on exclusive rights-of-way and penetrates the densest and most congested areas. It provides access of the highest quality, while serving major generators and intercepting other transit facilities. The DPM belongs to and is an integral part of the transit family of modes. It is generally of the collector/distributor type but could be used for line-haul services. If applied properly, the DPM combined with a well designed line-haul mode is certain to present strong competition to the automobile.

Table 1. Qualitative assessment and comparison of transit functions

Facility function	Trip length	Elements linked	Trip components	Station spacing	Speed	Reliability	Capacity
Access	Very small	Individual sites; local areas	Collection/ distribution	Very short	Low	Poor	Small
Access/ movement	Inter- mediate	Secondary generators	Collection/ distribution and line-haul	Inter- mediate	Medium	Fair	Moderate
Movement	Long	CBD major generators	Line-haul	Long	High	Very good	Large

Table 2. Operational guidelines

Type of operation	Right-of-way category	Vehicle technology					
Local	Uncontrolled	Para transit	Surface bus	Trolley bus	Streetcar	--	PRT
	Partially controlled	Para transit	Bus partially on bus-way	Trolley bus	Light rail	Heavy rail	PRT
	Fully controlled	Para transit	Bus on bus-way	Trolley bus	Light rail	Heavy rail	DPM
Limited	Uncontrolled	Para transit	Surface bus	Trolley bus	Streetcar	--	PRT
	Partially controlled	Para transit	Bus partially on bus-way	Trolley bus	Light rail	Heavy rail	PRT
	Fully controlled	Para transit	Bus on bus-way	Trolley bus	Light rail	Heavy rail	DPM
Express	Uncontrolled	Para transit	Surface bus	Trolley bus	Streetcar	--	PRT
	Partially controlled	Para transit	Bus partially on bus-way	Trolley bus	Light rail	Heavy rail	PRT
	Fully controlled	Para transit	Bus on bus-way	Trolley bus	Light rail	Heavy rail	DPM

FURTHER READING

American Association of State Highway Officials, A policy on design of urban highways and arterial streets, 1973.

American Public Transit Association, Para-transit in the family of transit services, prepared by the Task Force on Para-Transit, 1976.

Bers, Eric L., "Busway versus rail transit - a land-use perspective", Transportation Engineering Journal of the American Society of Civil Engineers, September 1977. Institute of Transportation Engineers, Transportation and traffic engineering handbook 1977.

Morlok, E.K., "The comparison of transport technologies", Highway Research Record 238, 1968.

Myers, Edward T., "Matching the Modes", Modern Railroads, March 1976.

Vuchic, V.R., and Stanger, R.M., "Lindenwold rail line and Shirley busway: a comparison", Highway Research Record 459, 1973.

Vuchic, V.R., "Place of light rail in the family of transit modes", presented at the National Conference on Light Rail Transit, Philadelphia, 1975.

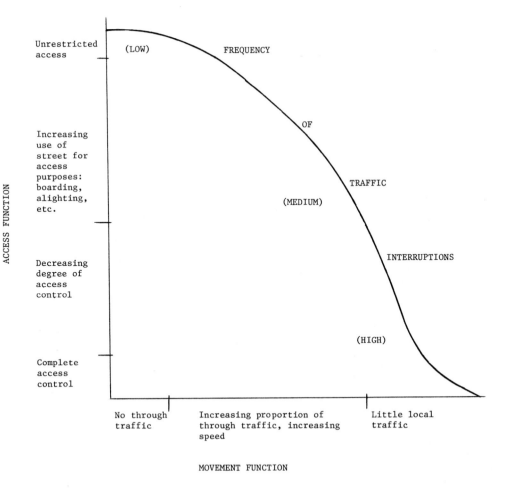

Figure 1. Transit function - movement versus access

A rational approach to trip generation   12

Richard J. Brown   South Africa

SYNOPSIS

The trip generation phase of transportation planning is the first stage and thus the foundation of the travel demand estimation process. It is the essential link between land use and planning and the derived demand for travel. Many of the problems we face are due to the separation of land use and transportation into two separate and almost closed systems. An understanding of travel demands that all the functional aspects of land-use development, including travel generation, be understood as a single system. To date, the estimation of travel demand has been based on two methods: multiple linear regression analysis, and category or cross-classification analysis. On close inspection, it is often found that many of the basic assumptions of these models do not hold when the methods are applied to travel demand estimation. The purpose of this paper is to illustrate some of these errors of assumption and to compare the two general methods. The estimating process must be structured in such a way that more flexibility can be introduced to the modelling process. More use should be made of pilot studies to determine important variables before decisions on the shape of the appropriate response surface are constrained by inappropriate survey data. The paper discusses factorial design experiments and the simple method of variable evaluation by the application of Yates' algorithm.

INTRODUCTION

Before discussing some of the problems associated with the trip generation stage of the urban transportation planning process and outlining some possible ways of improving present performance, it is necessary to define the role of this particular phase of the process. Trip generation is concerned with the choice of moving a location in pursuance of some personal activity with either is not possible at the present location or cannot be conducted with the same efficiency at the trip origin as it can be at the proposed trip destination. The choice that has to be identified and determined is thus primarily concerned with person activity patterns rather than other segments of the travel decision process of destination choice, mode choice and route choice.

Thus the basic concern is with those activities of a person or household that require a change in location for their successful execution. An understanding of the functional relationships involved can only be achieved with a systems approach. From the study of activity sets, not only the travel demand will emerge but also the land-use pattern which is most important in destination choice. The input variables that are of interest in determining trip generation must therefore reflect the functioning of activity systems rather than static variables of land use, such as population, car ownership, residential density and floor areas.

Such an approach will offer the opportunity of planning land use and transport as one integrated system from one common set of inputs. In the past much as been written and spoken of the need for such an integrated approach but this has not been realised in practice. Many of the problems associated with the urban planning process have arisen because the two major physical sub-systems, i.e. land use and transport, have been studied as separate issues. We have been faced with the problem of "and" as identified by Eddington in the planning approach.[1] This problem can be summarised as a situation which allows a detailed knowledge of each element of a compound without an understanding of the combination of these elements. For example, a deep knowledge of hydrogen (H) and oxygen (O) does not mean an understanding of the properties of water (H_2O). So it is with land use and transport planning. One hopes that activity systems will provide the key to that understanding.

The object of this paper is thus to discuss some of the implications of this approach to planning in the context of recent developments in both the behavioural and statistical aspects of the current approach to trip generation modelling.

BEHAVIOURAL ASPECTS

The current set of trip generation models have not changed significantly in behavioural content since their inception in the early studies (Chicago and Detroit) of the late 1950s. Chapin[2] at a conference on urban development models in 1967 was one of the first to put forward the concept of household activity systems, even though this related more specifically to land use rather than transport models.

The household activity system is conceived around a framework of choice theory connecting up the value system of a person with his activity system. Activities can be viewed as either discrete episodes in the life of a person, e.g. visiting the cinema, or in terms of routines. A person's or household's routine was defined by Chapin as a recurring sequence of episodes in a given unit of time, e.g. going to work. These routines are subject to a feedback system which can be identified on four time scales: the daily cycle, the weekly cycle, the seasonal cycle and the life cycle. Chapin concluded by stating that in terms of land-use patterns, households could be grouped by life-style and stage in the life cycle and that such groups would show consensus on a number of norms.

Cullen[3] extended this approach to study the constraints and difficulties arising from space and time disruptions experience by those operating in the city context. The study was concerned with the application of activity sets to land use rather than transport planning, however.

Perilla,[4] in a study of urban structure and residential choice, was one of the first to study basic activity patterns in the context of both urban and transportation planning. Perilla found that much of the travel decision-making process was predetermined by residential location decisions. The strongest motivating force in the latter decision among all socio-economic groups was a desire for an increase in living space. Among the professional and administrative groups, the convenience of transportation to work was only sixth in importance, while the convenience of transportation facilities and other services appeared to be a weak motivating force among all groups.

Another finding of Perilla's study which is important to transportation planning, was the apparent contradiction of Zipf's principle of least effort.[5] This principle postulates that people prefer the behaviour that requires the least expenditure of effort to reach their goal. It was found that those with the highest degree of choice, i.e. with least constraints, tend to live furthest away from all transport and civic amenities. This finding of course has important implications for the use of models, such as the gravity and other utility maximising models. This particular finding was recently confirmed by Todes in a study in Cape Town.[6] He concluded that the higher income groups tend to live further away from their place of work. Figure 1 shows a comparison between the work trip travel time distribution of the upper income group and that of other income groups.

More recently, direct relationships between activity systems and resultant transportation demand were studied by the Transport Studies Unit at Oxford University. Heggie[7] has summarised the TSU's modelling approach as socio-psychological models of travel choice, an approach which the conventionally trained transportation planner and engineer are not formally equipped to handle. Perhaps the time is rapid approaching when transportation will need to be taught as a complete subject so that the recipients of such training will emerge not as planners or engineers but as transportationists.

Following a series of empirical studies based on an in-depth interviewing technique, Heggie identified five aspects of activity behaviour which appear to be important influences on travel behaviour. These are summarised as follows:

□ Stage in the family life cycle. Six different stages have been identified within which travel behaviour changes markedly.

□ The incidence of spatio-temporal constraints. Certain activities can only be performed at certain times and places, e.g. shop opening hours. There are also a number of temporal constraints internal to the household, such as sleeping and meal times.

□ Capacity for adapting travel patterns to activity patterns. Households are immensely adaptive organisms and can achieve an objective in an infinite variety of ways.

□ Car as a joint asset. Not only is this true within each household, but also within neighbourhoods there is a significant degree of informal car pooling.

□ The psychology of choice. Factors such as habit, routinisation and limited knowledge of alternatives all appear to be important factors in the way the household travel decisions are formulated.

Conventional travel forecasting models are at present dominated by engineering and economic method and theory and tend to neglect the more important human behavioural elements outlined above. It emerges from this brief review of behavioural studies of travel demand that modelling procedures should attempt to reflect at least the following three factors:

□ stage in family life cycle (in perhaps six categories as suggested by Dix[8]);

□ family life style, to reflect the level of constraint imposed on travel behaviour by residential location using a socio-economic categorisation;

□ and transport availability, as opposed to car ownership, to select the transport actually available to the trip maker, rather than potentially available.

STATISTICAL ASPECTS

The statistical aspects of the travel demand procedure are concerned with model formulation and output accuracy. So far, the formulation of trip generation models has not advanced beyond the point of selecting a convenient computer package, often without a knowledge of the underlying principles of that particular form of model. Beyond presenting the results of such a model which has been calibrated with the same data, little has been done to assess the accuracy of the model output. In such circumstances high levels of agreement are to be expected.

Before selecting a model form, it is necessary to consider the role and the aims of that model. According to Heggie the model has to assist in understanding and explaining behaviour, assist with policy formulation, and provide accurate predictions for design and evaluation.[7]

It should be emphasised that the role is one of assistance to the decision maker. It does not act as a substitute for that function. Stopher and Meyburg[9] and Wilson[10] have put forward basic aims and check-lists to assist in model formulation. The three basic properties are simplicity (it should be as simple as possible and therefore intelligible to the non-specialist); utility (it should do the job for which it was designed); and validity (the output of the model should be valid, non-contradictory and generally applicable.) If these basic properties are met, the other desirable qualities are likely to be attained implicitly.

Accuracy is a combination of utility and validity. Alonso warned more than a decade ago of the dangers of error propagation in predictive models which can arise from the mathematical form of the model and the quality of data used in both the calibration and the predictive stages of the model.[11] More recently, Robbins reported that there was virtually no research into the overall accuracy of mathematical modelling.[12] He also showed that the 95 per cent confidence limit for a calibrated estimate of zonal generation was on average approximately 50 per cent. This estimate was based on a coefficient of variation of 100 per cent for individual trip rates in the category analysis model. The author has analysed other reported work with a similar model to discover that the coefficient of variation can vary between 100 and 700 per cent.[13] [14]

The modelling forms traditionally used for trip generation estimation have consisted of either multiple linear regression or category (or cross-classification) analysis. More recently, a considerable amount of literature has been produced on individual choice (or disaggregate behavioural) models.[15] [16] These models have been theoretically derived from basic economic and psychological theory and while to date they have been mainly applied to distribution and model choices, they are reputed to be able to handle trip generation decisions. However, in the context of the above discussion on behavioural aspects, there appear to be two major faults in the underlying assumptions.

The first is that the modelling procedure requires that the individual has a choice in his decision making but, as we have seen, the constraints imposed by routinisation and other spatial and temporal commitments often remove this element of choice. Secondly, in respect of residential location, it has been found that as the constraints are removed the resultant trip making is not subject to minimisation of effort or, in terms of the model, maximisation of utility. In summary, it can be said that in trip making there is often no choice, and where it does exist utility maximisation is not an appropriate consideration. This type of model form can therefore be discarded for the purposes of trip end estimation.

The multiple linear regression form, most extensively used in the past, has two major faults. First, it assumes that the regression surface is in fact linear, which often it is not. Secondly, the form of the data does not comply with the basic statistical assumptions on which the model is built. The non-independence of the independent variables and the use of the relationships for predictive purposes beyond the scope of the data used for calibrating are but two common examples of cases where this assumption is violated.

The category analysis form of the model has been described by Winsten as a very proper form of regression analysis that should be used by those with a feel for data.[17] It has the great advantages that it does not assume the shape of the regression surface, it handles disaggregate data and its form and conception are simple. However, it has been criticised by Douglas on the grounds that in its present form it does not produce the statistics by which its performance can be assessed.[14] Also, the present choice of variables results in a violation of the assumption that the between-category variance in trip rates is greater than the within-category variance.

It is considered that category analysis is the correct basic model form but that it needs further development to overcome its present shortcomings. The next section will deal with this and with the question of behavioural improvement, which can be considered as a rational approach to trip end estimation in the context of the present level of knowledge.

THE RATIONAL APPROACH

In suggesting improvements to the present approach to trip generation, it is very tempting to advise a complete revolution. This is unlikely to be well received because those in practice will not be prepared to discard overnight the knowledge and expertise built up during their professional careers. For any improvements to be readily accepted, therefore, it is necessary that they be suggested in an incremental fashion. The following discussion is aimed at introducing both statistical and behavioural improvements to the extensively used category analysis model.

At the same time, there is a need to introduce a greater level of flexibility into the modelling process. To achieve this, much more thought and effort are required at the design stage of the study. Usually, once financial approval is gained, researchers feverishly collect data, only to find later that the quality of data hoped for has not been achieved. At the data collection state there is also a tendency for reasons of convenience to concentrate on the household, with the result that non-home-based trips, especially those in the pedestrian mode, are not shown to be significant. But a visit to the CBD of any large city will show the huge amount of travel by this mode during the working day. Thus, not only do the behavioural and statistical content of the models need to be improved, but the whole approach to modelling needs to be reconsidered.

Development of the category analysis model. Current application of the category analysis model is confined to the determination of the mean trip rate for each category of combination of variables predetermined at the survey design stage. There are three consequences of this practice.

□ The effect of each of the variables and the variable interactions on the outcome (trip rate) is not checked.

□ The distribution of the within-category trip rates is not analysed to check how well the mean represents the observations. Low standard error of the mean is usually reported, but this is basically due to the high number of observations. As reported above, a limited investigation has indicated that the coefficient of variation can take on values of between 100 and 700 per cent. This latter finding indicates that the observation may not be normally distributed.

□ No knowledge is gained of the shape of the response surface of the model.

To overcome these and other statistical difficulties, a more rigorous statistical approach is proposed based on the experimental design methodology put forward by Guttman and others.[18] Experimental designs can be of two types:

□ Factorial designs, used when the objective is to screen "k" controlled variables in order to discover subsets whose effects on the response are greatest, without making assumptions about the shape of the response surface.

□ Response surface designs, used to meet a graduating objective in which a map of the function is drawn up over selected ranges of the controlled variables.

It is suggested that trip generation modelling should be conducted as a two-stage exercise. In stage one, a two-level factorial design is drawn up to represent all possible combinations of two versions of each of the "k" variables under consideration. This results in 2^k experiments, each with its own outcome. Certain statistics known as contrasts are computed, to assess the mean performance of each of the variables to determine whether there is a meaningful linear or quadratic trend associating the trip rate with combinations of variables.

A method known as Yates' algorithm greatly eases and speeds the computations required in analysing a two-level factorial design. This method gives the estimated effect of each of the variables, singularly or as interacting combinations, in terms of

$$E = \bar{y}_{k+} - \bar{y}_{k-}$$

where E = estimated effect;

$\bar{y}_{k+}$ = mean of the trip rates in those instances where variable k is set high; and

$\bar{y}_{k-}$ = mean of the trip rates in those instances where variable k is set low.

From this method it is also possible to compute the sum of squares attributable to each variable and each interacting combination, and therefore to draw up an analysis of variance table, establishing the significance of each of the effects.

Having thus established which of the variables have significant effects, one proceeds to the second stage, in which a response surface is constructed in terms of those particular variables.

Such surfaces may be:

1st order (linear) which take the form

$$y = a_0 + a_1 x_1 + a_2 x_2 \ldots\ldots\ldots\ldots a_k x_k$$

$$= a_0 + \sum_i a_i x_i$$

or   2nd order (quadratic) which take the form

$$y = a_0 + \sum_i a_i x_i + \sum_i a_{ii} x_i^2 + \sum\sum_{ij} B_{ij} x_i x_j$$

It is also possible to go to a third order model which involves contoured surfaces, as opposed to contour lines used in first and second order models. The theory behind the modelling process permits the computation of analysis of variance tables to identify the significant of the first and second order coefficients, i.e. those due to the regression as well as Lack of Fit and Error sum of squares.

In this particular field it may happen that certain variables under consideration cannot be quantified. In that case it will be necessary to restrict tha analysis to a multi-level factorial design in which the effect of each of the variables is assessed, rather than the formation of a fully graduated response surface. Chatterjee and Khasnabis[19] and Dobson[20] in recent papers have outlined the application and relevance of these techniques to transportation planning.

Variable selection. Since Woolton and Pick[21] put forward the category analysis method in 1967, there has been little change in the variables used, i.e. household size, income and level of car ownership. The only slight variation in studies employing this technique has been in the definition of category boundaries. There appears to be a case for reviewing the effectiveness of these variables against those suggested above. As recently reported by Williamson and Cooper, in the Brighton (UK) study income did not appear to be significant in trip making.[22] The author has carried out some preliminary analysis of the Johannesburg study data with Yates' algorithm. Here too, income did not appear to be effective, except in interacting combination with the number of non-employed persons per household with respect to home-based other trips (i.e. not work and school trips).

The author intends to carry out research with these model-building techniques to test the effectiveness of the conventional variables against the following suggested variables: stage in the life cycle; life-style; and transport availability.

The importance of the stage in the life cycle as a variable relevant to trip making has been identified by a number of researchers, and in the first instance it is intended to use six categories suggested by Dix.[8] These are:
□ young adults, married or not, without children;
□ families with dependent children of whom the youngest is aged seven years or less;
□ families with dependent children of whom the youngest is aged twelve years or less;
□ families with dependent children of whom the youngest is aged 13 years or more;
□ family of adults, all of working age;
□ elderly people.

To reflect the life-style of the household, it is intended to use a socio-economic grouping divided into nine categories as follows:

Income	Social group
High	Professional
Medium	Skilled
Low	Unskilled

In order to reflect the informal pooling of the use of cars outside the household it is intended to categorise transport availability as follows:

Number of family vehicles available	Neighbour's vehicle available for use	Public transport available
0	Yes	Yes
1	No	No
2+		

This would give twelve categories of transport availability.

Combining the three variables described would result in 648 categories which might be considered excessive, especially in view of the amount of data required to give reliable means for each of the categories. If, however, this set of variables were to lead to mean trip rates for which the within-category variance was but a small percentage of the mean, much smaller sample sizes would be acceptable. For example, if the coefficient of variation was reduced from 100+ per cent to about 10 per cent, sample sizes as low as 10 would give a reliable estimate of the mean.

While both transport availability and life-style are suitable for a two-level factorial analysis, the variable "stage in the life cycle" is not, unless various combinations of pairs of stages are used in the preliminary analysis. The use of these variables in the final analysis would necessitate the adoption of the multi-level factorial approach as it would not be possible to apply regression techniques because of the non-quantitative nature of the variables used.

Practical considerations. In putting forward suggestions for a more rational approach to trip generation modelling, it must be recognised that we do not have much knowledge of travel demand generation. Therefore, we must approach the problem in the same way in which the industrial scientist uses a properly constructed experimental design to approach the research problem. At the outset, we must undertake a pilot study with a two-level design which will enable the planning team to understand better the issues involved before commencing the major study.

A second area of concern is the almost complete reliance on the home interview study for all journey data. While it is recognised that home-based trips are fairly reliably reported, non-home-based trip data require considerably more attention. The author was involved in a study in the UK which was based on vehicle trips. A screen-line check found that the expanded survey data produced only four per cent of non-home-based business trips recorded at the screen-line. The reporting of non-home-based walking trips is probably even less reliable. It is suggested that far greater effort should be made to collect data at locations where high generators are involved. The present inadequate facilities for pedestrians in the CBD may be a direct reflection of the lack of data produced by the normal study on such movements. The study design needs to be made more flexible so that the data collection for various trip purposes and travel modes is designed in such a way as to provide the most efficient and effective data for that particular category of trip.

CONCLUSIONS

Some of the problems that have to be faced in the trip generation modelling process have been presented and some suggestions for overcoming these difficulties put forward. The suggestions are incremental in nature in that they can be grafted on to the present methodology. It should be emphasised, however, that these proposals are the result of a synthesis of recent research findings in a number of different fields and at this stage must be considered no more than a hypothesis.

REFERENCES

1 Eddington, Sir A., The nature of the physical world (1958).
2 Chapin, F.S., Activity systems as a source of inputs for land use models (1968), Highway Research Board Special Report 97.
3 Cullen, I.G., "Space, time and the disruption of behaviour in cities" (1972), Environment and Planning Vol 4.
4 Perilla, O., "Motivation limiting principles, household characteristics, urban structure and residential choices" (1972), Journal of Environmental Systems Vol 2(1).
5 Zipf, G.K., Human behaviour and the principles of least effort (1949).
6 Todes, M.A., A regression technique for the calibration of a singly constrained gravity model (1978), MSc thesis, University of Cape Town.
7 Heggie, I.G., "Socio-psychological models of travel choice - the TSU approach" (1977), Traffic Engineering and Control.
8 Dix, M.C., Report on investigations of household travel decision-making behaviour (1977), Working Paper 27, Transport Studies Unit, Oxford University.
9 Stopher, P., and Meyburg, A., Urban transportation modelling and planning (1975).
10 Wilson, A.G., Urban and regional models in geographing and planning (1974).
11 Alonso, W., The quality of data and choice and design of predictive models (1968), Highway Research Board Special Report 97.
12 Robbins, J., "Mathematical modelling - the error of our ways" (1978), Traffic Engineering and Control.
13 Chatterjee, A., Martinson, D., and Sinha, K., "Trip generation analysis for regional studies" (1977) Transportation Engineering Journal of ASCE Vol 103 (TE6).
14 Douglas, A., "Home-based trip end models - a comparison between category analysis and regression analysis procedures" (1973), Transportation Vol 2.
15 Richards, M., and Ben-Akiva, M., A disaggregate travel demand model (1975).
16 Stopher, P., and Meyburg, A., Behavioural travel demand models (1976).
17 Winsten, C., Regression analysis versus category analysis (1967), Proceedings of PTRC Seminar: Trip End Estimation.
18 Guttman, I., Wilks, S., and Hunter, J., Introductory engineering statistics (1971).
19 Chatterjee, A., and Khasnabis, S., "Category models - a case for factorial analysis" (1973), Traffic Engineering Vol 44(1).
20 Dobson, R., "The general linear model analysis of variance: its relevance to transportation planning and research" (1976), Socio-Economic Planning Sciences Vol 10.
21 Wootton, J., and Pick, G., "Trips generated by households" (1967), Journal of Transport Economics and Policy Vol 1(2).
22 Williamson, J., and Cooper, M., "The Greater Brighton transportation study trip generation model" (1977), Traffic Engineering and Control.

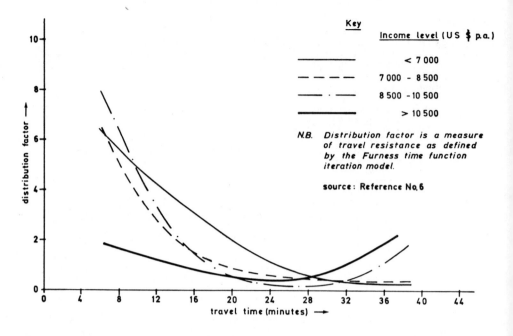

Figure 1. Variation in effect of travel time on distribution factor with changes in income level

The effect of income on trip-making behaviour

13

Erella Daor   Israel

SYNOPSIS

The paper presents some of the findings of a study on the effect of income on trip-making behaviour. Two main aspects were examined: the effect of income on trip frequency, i.e. trip generation; and the effect of income on trip length, cost and time distributions. Many variables are likely to affect the number of trips generated, but while it is valuable to include all relevant independent variables and thus reduce specification errors, it is doubtful that it would be advantageous to do so when, as in the case of transportation studies, the input data are not always reliable. In addition, some variables are inherently difficult to forecast, and the possible gain from their inclusion might be neutralised by the errors involved in forecasting their future values.

In the trip generation model developed for the Greater London Transportation Study (GLTS), variables hypothesised as having a causal effect on trip generation were tested for significance, and their contribution to the explanation of variation examined. The model was calibrated and tested using data collected in a home interview survey conducted in the period September 1971 to June 1972 in the London area. Income is considered to be a major socio-economic variable. When tested, income was found to be significantly correlated with both work trips and non-work trips, but its contribution to the explanation of variation was less than one per cent when the effects of other variables were accounted for. As this was contrary to expectation, a further analysis of the effect of income on other aspects of trip making, such as trip time, cost and length, was undertaken.

The distributions of trip length, cost and time by mode (private car and public transport) and by income (low, medium and high) of trips to work and other home-based trips (OHB) were calculated and compared. The means of the distributions of work trips for low, medium and high-income categories did not differ significantly. The same applied to OHB trips, except for the distribution of trip length by public transport, in which case the means of the distributions for the three income categories were significantly different.

The main finding of the research was that the effect of income on trip making, apart from its effect on car ownership levels, was small. Reliable estimates of trends in income distribution for long periods of time are hard to acquire, especially for small areas. Because income is a significant variable, however, further research would be necessary to discern its effect on travel choice. In particular, new data on travel after the introduction of high petrol prices might indicate whether the low sensitivity of trip generation in 1971 was mainly due to the low levels of petrol prices.

INTRODUCTION

Many variables are likely to affect the number of trips generated. According to economic consumer theory, the demand for goods depends on the characteristics of the consumer, including his income and tastes, and on the price and quality of the goods required. The variables likely to affect trip generation are of two distinct types: socio-economic and spatial. The socio-economic variables represent both the characteristics of the transportation system and the type and intensity of land use. In this paper we look at the former, which is a measure of the propensity to generate trips. We focus the analysis on income which is considered to be a major socio-economic variable, as it reflects the level of economic resources available to the household.

The findings in this paper are based on data collected for the Greater London Transportation Study (GLTS). The trip and socio-economic data were collected in a home interview study, one in a series of surveys conducted in London in the period September 1971 to June 1972.[1] In each interview information was gathered on the household, the members and their travels on the previous day. A household information tape was made which contains information extracted from the basic home interview records. It records the number of trips of various kinds made on a weekday, together with demographic data for each household.

The research was carried out as part of the development of a trip generation model for the GLTS strategic model. At the trip generation stage, variables hypothesised as having a causal effect on trip generation were tested for significance, and their contribution to the explanation of variation examined.

Transportation studies, based on category analysis, have included income as one of the explanatory variables.[2] [3] Unfortunately, income is inherently difficult to forecast, especially for periods of over five years and for small area units. If income data were needed at traffic zone level, many assumptions about housing stock and social changes would have to be made. For these reasons, the effect of income on trip-making behaviour was closely examined.

The analysis was carried out at the household level and applied to home-based trips. Thus, the term "trip" is used here for home-based trips, unless otherwise defined. It was carried out separately for work and non-work trips, as the nature of the two trip types differs substantially. Work trips are marked by their inelasticity in terms of frequency, time and location, while non-work trips are more flexible in terms of frequency, choice of destination and time. The contribution of income to the explanation of variation in the number of trips generated was studied, together with other socio-economic variables, such as the number of employed members of the household, car ownership levels and household size.

The effects of socio-economic variables are usually non-linear, and it is therefore unlikely that they will be of much help as quantitative variables in a standard regression analysis. Instead, each variable is treated as a factor with a fixed number of levels. The cross-classification of the trip rates by each factor as defined by its levels, defines a number of categories, each with a distinct trip rate. The data were analysed by a non-orthogonal analysis of variance programme which systematically builds up a factorial model step by step.[4]

The process demands fitting the following succession of models to the data:

(1) $Y_{ijk} = \mu + e_{ijk}$

(2) $Y_{ijk} = \mu + \alpha_i + e_{ijk}$

(3) $Y_{ijk} = \mu + \alpha_i + \beta_j + e_{ijk}$

(4) $Y_{ijk} = \mu + \alpha_i + \beta_j + \Upsilon_k + e_{ijk}$

The α_i are the parameters of the first factor to be brought into the model equation (i.e. the one whose contribution to the explanation of variation is the highest); the β_j and Υ_k are the parameters of the second and third factor respectively.

To enter the model, the variables had to be significant at the 1 per cent level, and to contribute more than 1 per cent to the explanation of variation, i.e. the added R^2 should be greater than 0.01.

The second requirement prevented some of the variables from entering the model solution. While the introduction of more parameters will reduce the error, this may be at the expense of the robustness of the model. For example, a two-factor model may be more satisfactory than one with a third factor, even though the latter will explain more of the variation. Likewise, a ten-way breakdown of income may reduce the error more than a three-way breakdown, but we will then have to carry forward ten parameters for a prediction instead of three. The criterion here is partly one of simplicity and partly one of statistical significance of the estimated parameters. The possible loss of explanatory power is quite small, as the model retains the factors which add more than 1 per cent to the explanation of variation.

Income was found to be significantly correlated with both work and non-work trips, but its contribution to the explanation of variation was less than 1 per cent when the effects of other variables were accounted for. The analysis is described below.

WORK TRIPS

The dependent variable was taken to be the number of daily trips from home to work generated by a household. The independent factors with their corresponding levels were the number of employed residents (1,2,3+), income (low, medium, high), and car owernship (0,1,2+). The added R^2s (explanation of variation) of each of the factors on its own was: employed residents, 0.310; income, 0.085; and car ownership levels, 0.042.

The first factor to enter the model equation was the number of employed residents. Since its effect was accounted for, the R^2 values of the other two factors were: car ownership, 0.011; and income, 0.009. Thus, car ownership was the second factor to enter the model equation. When the effects of both the number of employed residents and car ownership levels were accounted for, the added R^2 due to income was only 0.003; thus it did not enter the model equation.

It is interesting to note that while income on its own gave a higher explanation of variation than car ownership, once the contribution of the number of employed residents was accounted for, the added explanation of income was smaller than that of car ownership. The reason for this is that the factor of employed residents is more closely associated with income than with car ownership; therefore, the added contribution of income after accounting for employed residents is smaller. The finding shows quite clearly that although the constraints imposed on the inclusion of variables in the model equation had prevented income from being included, the loss in explanatory power was very small (0.003).

OTHER HOME-BASED (OHB) TRIPS

A similar picture emerged when the effect of income on OHB trips was studied. The dependent variable was taken to be the number of daily (week-day) non-work trips from home generated by a household. The independent factors and their corresponding levels were household size (1,2,3,4,5+), car ownership (0,1,2+), number of employed residents (0,1,2+), and income (low, medium, high). The correlations between OHB trips and the socio-economic factors were as follows: household size 0.192; car ownership 0.158; income 0.074; and employed residents 0.033.

The association between OHB trips and household size and car ownership is much stronger than the association between OHB trips and income. After accounting for household size and car ownership levels, the contribution of the other factors was very small. Income adds only 0.007 to the explanation of variation.

The group of OHB trips is quite heterogeneous. It includes education trips which are similar to work trips in their regularity, weight in the individual daily schedule, and importance. It also includes trips for entertainment, which are generally infrequent, and trips to hospital, which can be expected to be fairly insensitive. It is possible that trip generation for different purposes would be affected by different factors. OHB trips were therefore disaggregated by purpose, and the analysis was carried out separately for the following purposes: shopping and personal business, social, sport and entertainment, and education. The dependent variables were defined as the number of trips from home per household on an average weekday for each of these purposes. The independent factors were the same as before. The factors which entered the model equation for each trip purpose are summarised in Table 1.

As can be seen in the table, income did not enter any of the model equations. Thus, even when OHB trips were disaggregated by purpose, the explanation of variation owing to income was less than 1 per cent, once the other factors were accounted for.

To investigate further the effect of income on OHB trip generation, the same analysis procedure was applied to data collected in 1962 for the London Transportation Study (LTS). The LTS trip generation model, based on 1962 survey data, included household income as one of the independent factors. The two surveys (1962 and 1971) were not identical, for changes in the definition of some of the variables were incorporated in the 1971 survey, which also has a completely different zoning system.[5] Therefore, in order to do a detailed comparison between the two surveys the data would have had to be put on a common basis. This was outside the scope of the present study, but it was nevertheless possible to apply to the 1962 data the same analysis of variance procedure which was used with the 1971 data. Using the 1962 data, a run was carried out to test the effect of income on OHB trip generation. Four factors were included in the run: car ownership, household size, number of employed residents, and income. Income was classified into three levels: low, medium and high. The classification took account of the growth in income from 1962 to 1971.

Car ownership was the first factor to enter the model equation with an R^2 value of 0.07. The second factor was household size which added another 0.04 to the explanation of variation. Income was the third factor which added an R^2 value of 0.024. The number of employed residents and the interaction terms, although significantly correlated with OHB trips, contributed less than 1 per cent to the explanation of variation. These results show that the contribution of income to the explanation of variation in OHB trip generation was greater in 1962 than in 1971.

OTHER ASPECTS OF TRIP MAKING

The foregoing analysis shows that the effect of income on the number of trips generated from home is small. The question arises whether the effect of income is more pronounced for other aspects of trip making. For instance, do households with high incomes undertake longer and costlier trips than those with low incomes? People with high incomes may be more selective in their place of residence and work. They may choose to live in a secluded suburb (e.g. the stockbroker belt), and commute over longer distances to work. They may also have a wider choice in their selection of destinations of OHB trips. A shopping trip can be made to the nearby grocery store or to the distant shopping centre; and a trip for entertainment can be made to the local pub or to an exclusive club.

To find out whether the effect of income is manifested in the type of trips undertaken (i.e. trip length, time and cost), the trips were classified according to the income of the household. Work and OHB trips were classified according to three income levels: low, medium and high. For each trip purpose and income class, five distributions were calculated: the distributions of trip length by private car and public transport; the distributions of trip time by mode, and the distribution of trips by trip cost by public transport. The cost of a trip by private car was not calculated as it depends on the type of car (make, engine size and age), and the data were not available. The trip length, time and cost were calculated as follows:

□ The length of a trip was defined as the sum of the component stage distances. The length of each stage was defined as the crow-fly distance between the centoids of the zones of origin and destination of the stage. If both the origin and destination of a stage lay in the same traffic zone, an intra-zonal distance was assigned to the stage. (A full list of intra-zonal distances is given in an Appendix; see page 140).

□ The trip time was calculated by subtracting the origin time of the first stage of the trip from the arrival time of the last stage of the trip.

□ The cost was defined as the sum of the component stage costs as reported in the survey. Where two consecutive stages had the same ticket validity, the stage cost was counted only once.

The averages of trip length, time and cost distribution of work trips by mode and income class are given in Table 2. A complex pattern emerges from this table. In trips made by private car, the trend is that, on average, the lower the income, the longer (both in distance and time) the trips. The reverse is true for public transport: the higher the level of income, the longer (in length and time) the trip. People of middle income make the shortest and least expensive trips. The question arises whether the observed differences in the means of the three income classes stem from real difference in behaviour, or whether they can be attributed to chance variations or measurement errors. A mean test applied to the data showed that the difference in trip length by public transport between medium and high-income groups was significant at the 5 per cent level. The differences between the other means, however, were not statistically significant. (The mean test statistics for work trips distributions are given in the Appendix.)

Table 3 gives the averages of OHB trip length, time and cost distribution by income class. This table shows quite a simple pattern. People with high incomes make longer and costlier trips than those with medium incomes, while the latter make longer and costlier trips than those with low incomes. The mean test statistics for OHB trip distributions were not significantly different, except for trip length by public transport. The means of the three distributions (trip length by public transport for those with low, medium and high incomes) were significantly different at the 5 per cent level. (The results of the means test for all the distributions are given in the Appendix.)

CONCLUSIONS
On the basis of 1971 data, income did not contribute significantly to the explanation of the variation in the generation of either work trips or OHB trips. Even when the latter were disaggregated by purpose, the added explanation due to income was less than 1 per cent when the contributions of the other factors were accounted for. When the same analysis was applied to 1962 data on OHB trips, income was introduced into the model equation and the R^2 added for its inclusion was 0.024. The difference in the significance of the income effect may be due to the relative cheapness of travel in 1971. Once a household acquired a car, its income did not affect the number of trips made. The analysis of trip cost, length and time distribution has shown that the most significant deviations are those between the trip length distributions by public transport of the low income group and the high income group.

The effect of income was found to be more significant in the 1962 data. This leads us directly to the following question on income. Are there any deficiencies in its evaluation? Income information is hard to acquire and its accuracy is always suspect. At the end of the interviews in the GLTS survey households were asked to classify themselves into one of ten income levels. In 18.5 per cent of otherwise successful interviews, income data were missing, and in order not to reduce the sample size, income data were "patched" by estimates from two other control variables, namely the socio-economic group (SEG) of the head of the household, and the number of employed residents (RM 310). The implied assumption of this method is that households which omitted to answer questions have incomes similar to those of other households with the same values for the variables - size and number of employed residents. To date, there is no evidence to the contrary, and it is reasonable to assume that the data on the whole are quite reliable. Moreover, income data were collected in ten categories, but in this research a three-way split was used (low, up to £1,500 a year; medium, up to £3,000; and high, over £3,000). Therefore the error is probably small within this broad classification. We can conclude, therefore, that at least in the 1971 situation, when petrol was relatively cheap, the effect of income on the number of trips was small, over and above its effect on car ownership.

The main finding of the research was that the effect of income on different aspects of trip-making behaviour (i.e. frequency, length, time and cost) is small. This conclusion is limited in scope because the activities of trip making in London may not be valid for other localities. Nevertheless, a recent study on travel patterns in urban areas in Israel showed that the effect of income on overall mobility was very small.[6] Income elasticities were positive, but very low.

The results of this research preceded the energy crisis. As income is a significant variable, further research would be necessary to discern its effect on travel choice. In particular, new data on the frequency of travel after the escalation in petrol prices may indicate whether the low sensitivity of trip generation in 1971 was mainly due to the low levels of petrol prices. The minor effect of income is quite welcome from a prediction point of view. Reliable estimates of trends in income distribution for long periods are difficult to obtain, especially for areas smaller than the GLC - for instance, zones or boroughs. Though income is the main ingredient of forecasts of car ownership levels, it has been found that projected levels of car ownership do in fact occur around the year of the forecast.

Table 1. Factors affecting OHB trips by purpose

Trip type	Household model
Shopping and personal business	Car ownership Household size Employed residents
Sports and entertainment	Car ownership
Sport	Car ownership
Education	Household size

Table 2. Averages of trip length, time and cost distribution of work trips by mode and income class

	Income class		
Average	Low	Medium	High
Trip length by car	7.0 km	6.9 km	6.3 km
Trip time by car	25 min.	25 min.	22 min.
Trip length by public transport	10.1 km	9.7 km	11.2 km
Trip time by public transport	46 min.	46 min.	47.5 min.
Trip cost by public transport	11.8 pence	10.9 pence	12.6 pence

Table 3. Averages of OHB trip length, time and cost distribution by income class

	Income class		
Average	Low	Medium	High
Trip length by car	6.1 km	6.4 km	6.8 km
Trip time by car	21 min.	20 min.	20.5 min.
Trip length by public transport	7.2 km	8.9 km	11.0 km
Trip time by public transport	42 min.	43 min.	46.5 min.
Trip cost by public transport	7.1 pence	7.7 pence	10.1 pence

REFERENCES

1   Stroud, A.A., GLTS : Initial processing of household interview data (1974),
    Research Memorandum 310, Greater London Council.
2   Wootton, H.J., and Pick, G.W., "A model for trips generated by households" (1967),
    Journal of Transport Economics & Policy, Vol 1, pages 137-53.
3   SELNEC Transport Model (1970), MAU-N-200.
4   Daor, E., The estimation of travel demand (1976), Research Memorandum 490,
    Greater London Council.
5   Crawford, Greater London Transportation Survey internal zone coding (1974), Research
    Memorandum 300, Greater London Council
6   Reichman, S., Instrumental and life-style aspects of urban travel behaviour
    (1977), Transportation Research Record 649, Transportation Research Board,
    Washington.

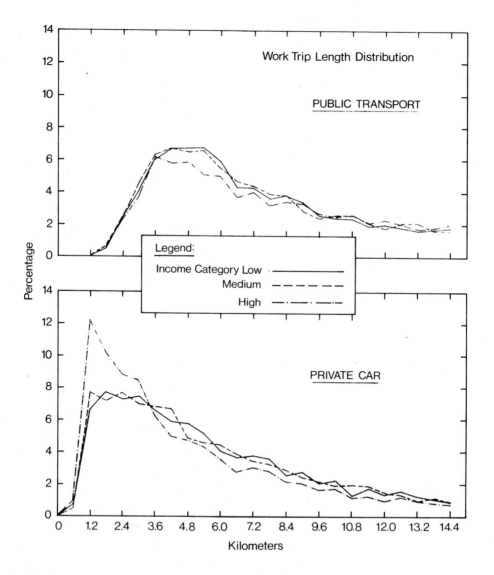

Figure 1

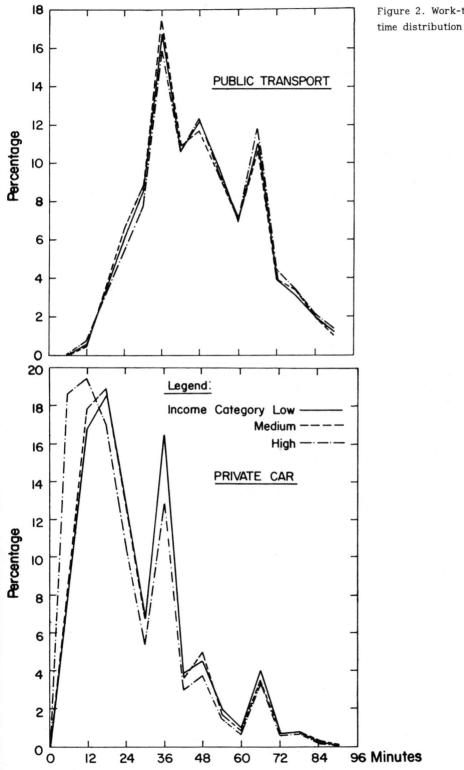

Figure 2. Work-trip
time distribution

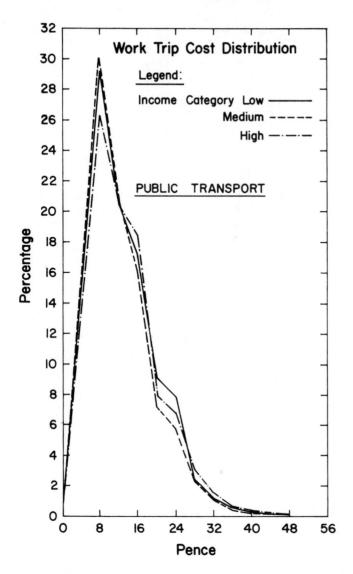

Figure 3

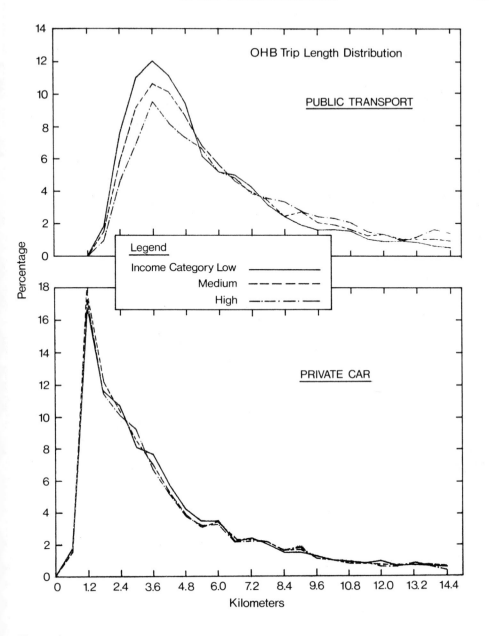

Figure 4

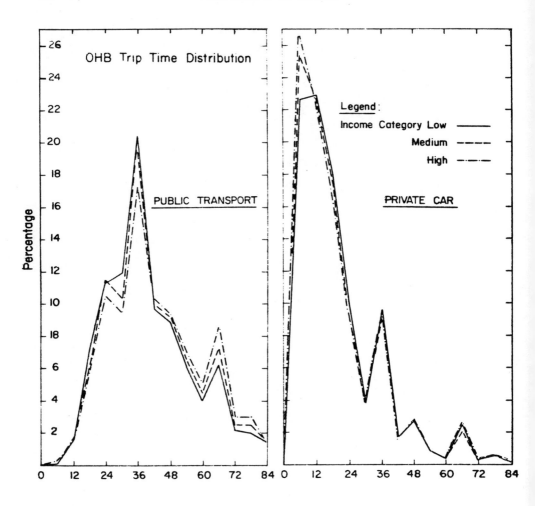

Figure 5

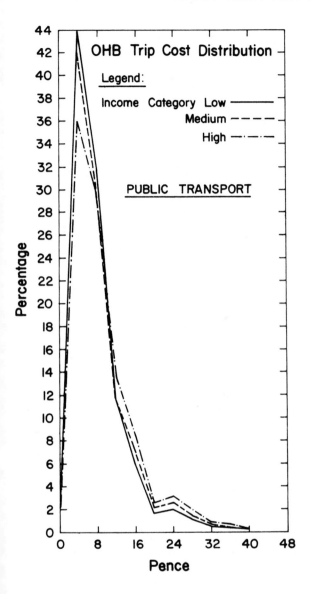

Figure 6

APPENDIX: STATISTICAL TESTS

A means test was applied to the distributions of work and OHB trips by income class. The results of the tests are given below. Distribution 1 refers to the low-income category, Distribution 2 to the medium-income category, and Distribution 3 to the high-income category.

Work trips

□ Comparison of trip length by private car: Distributions 1 and 2, 0.0412; Distributions 1 and 3, 1.2607; and Distributions 2 and 3, 1.2137.

□ Comparison of trip length by public transport: Distributions 1 and 2, 0.5350; Distributions 1 and 3, -1.4070; and Distributions 2 and 3, -1.9527

□ Comparison of trip cost by public transport: Distributions 1 and 2, 0.1594; Distributions 1 and 3, -0.5665; and Distributions 2 and 3, -0.7269.

□ Comparison of trip time by private car: Distributions 1 and 2, 0.0919; Distributions 1 and 3, 1.0751; and Distributions 2 and 3, 0.9871.

□ Comparison of trip time by public transport: Distributions 1 and 2, 0.2229; Distributions 1 and 3, -1.3949; and Distributions 2 and 3, -0.6154,

Only the means test statistic of trip length by public transport of medium and high-income categories was found to be significant at the 5 per cent level. Thus, the means of the distributions for low, medium and high-income categories (with one exception) of work trips are statistically not significantly different.

OHB trips

□ Comparison of trip length by private car: Distributions 1 and 2, -0.2043; Distributions 1 and 3, -0.3947; and Distributions 2 and 3, -0.1915.

□ Comparison of trip length by public transport: Distributions 1 and 2, -1.6724; Distributions 1 and 3, -3.6075; and Distributions 2 and 3, -19614.

□ Comparison of trip cost by public transport: Distributions 1 and 2, -0.4023; Distributions 1 and 3, -1.4023; and Distributions 2 and 3, -0.8422.

□ Comparison of trip time by private car: Distributions 1 and 2, 0.2177; Distributions 1 and 3, 0.1337; and Distributions 2 and 3, -0.0831.

□ Comparison of trip time by public transport: Distributions 1 and 2, -0.2987; Distributions 1 and 3, -1.0815; and Distributions 2 and 3, -0.7892.

The means test statistic is significant for OHM trip lengths by public transport. The means of the three income categories are significantly different.

Analysing transportation problems through daily life patterns

14

Velibor Vidakovic   Netherlands

SYNOPSIS

This paper reports on a travel/activity research programme carried out in the Netherlands. The subject of the study is the nature and structure of the transportation problem as part of the temporal and spatial organisation of daily life in an urban area. By an integrated analysis of both travel and non-travel components of daily life, the transportation problem is defined in terms of the generation of different classes in situations with regard to the activity linkages. The method applied in this study is a disaggregate observation, analysis and modelling of individuals' travel behaviour under the choice/constraint conditions created by social, temporal and spatial variables. The results show how the individual's daily life cycle (daily path) can be recognised as a basic unit of urban process and applied to the analysis of the transportation problem. A number of issues illustrate this approach, such as the individual's activity schedules as determinants of travel potential; the joint influence of home and work locations on the remainder of activities; the configuration of travel task as derived from the activity sequences; and the temporal and spatial interaction between transportation and activities. The conclusions refer to the research and planning concepts of the urban land use/transportation systems. The redefined mobility and accessibility concepts should embody the activity variables. It is suggested that a new criterion, transitiveness - that is, the degree in which the systems stimulate the transactions from one activity to another - should be applied to the evaluation of alternative land use/transportation networks. The significance is highlighted by two alternative urban structure patterns, where parallel and serial activity merge.

THE INDIVIDUAL'S DAILY PATH

Looking beyond the boundaries of the traffic scene enables us to derive the transportation task from the generalised task or urban organisation. That generalised task is the distribution of access to activities among individuals. The access to activities may be considered one of the basic measures of the quality of life. It expresses the degree of participation in the entire socially created spectrum of opportunities. While the distributional effects of any specific socio-economic organisation are commonly measured by the access to economic and cultural resources, the access to activities concerns a more fundamental human resource, time.

It is this close connection between the social order and the space-time order which makes it necessary to consider the transportation problem in a wider context of activity systems. The travel pattern interacts with the activity pattern. In any given stage of this interaction, the need for travel is determined in terms of time, space and volume by the way in which the activities of population are ordered. Therefore, to improve our understanding of the transportation problem we have to trace it back to the conditions which generate a given pattern in which people move around in their daily life.

Daily life in an urban area consists of a great number of activities and trips performed by the inhabitants according to a set of individual choices and collective constraints. Part of this pattern is travelling, which for some purposes can be observed separately. Within the context of the integration of traffic and transportation engineering in urban planning, however, we should analyse the transportation process in its close relationship to urban, socio-economic, temporal and land use activities.

It is obvious that the number of consecutive trips of an individual is only part of his history, containing information on his various social, spatial and temporal states. In operating his daily schedule, an individual converts his needs, subject to social and physical constraints, into the visible pattern of activities and trips. In any given socio-physical state on his daily trajectory, he processes a certain amount of information on both preceding and subsequent states. We suggest that an integrated urban transportation study should be able to describe, explain and evaluate the travel pattern as part of this sequential pattern.

Three requirements should be met before a transportation study can meet this general objective. First, we have to observe all trips including short walking trips. Secondly, we need information on activities that are the background of these trips. Thirdly, besides the conventional data processing, we should analyse the trips and activities within their individual order. As shown in Table 1, these three extensions of a conventional transportation study are considered essential for the analysis and interpretation of both travel need and transportation process.

For such an extension of the transportation study, a method of observation is needed to provide a detailed record of the individual's use of time and space during a consistent period. Specifically, the technique of the time and space budget survey meets this requirement.[1] This method involves a diary in which the sample persons record data on every single activity and every single trip on the given day.

For the purpose of this disaggregate approach, we consider an individual sequence of activities and trips during the shortest cycle, twenty-four hours, as a basic unit of transportation analysis. This unit is denoted by the term "individual's daily path", which is fully specified by the characteristics of all its serial elements as listed in Table 2.

An individual's daily path embodies a series of states, each specified by time, space and mode or socio-economic conditions. An episode (e) is an element of the individual's activity series and is devoted to one single activity (a). The successive episodes and activity transitions within the same location (address) constitute a sojourn (S), and a trip (T) is defined as a transition between two consecutive sojourns.

Thus, the simplest notation of an individual's daily path is given by a series:

$$S_1 (e_1 \ldots\ldots e_k) T_1 S_2 (\ldots) \ldots\ldots T_{n-1} S_n (\ldots),$$

S_1 and S_n being usually at home (h). By replacing the symbols of elements by their functional, temporal and spatial characteristics, the notation becomes:

$$p_o u_h (a_i t_1 \ldots\ldots a_p t_k) r_{ox} (m_q t_1) p_x u_s (\ldots) \ldots p_o u_h (\ldots a_m t_n).$$

Table 3 lists the basic variables in the study of the individual's path. The terms are explained in the Appendix (page 156). The daily path of an individual is a complex process, influenced by both long-run and short-run conditions of the individual's life. The general assumption is that an individual seeks the optimisation of his generalised time resources within the given constraints.[2] In the long run, individual choices can result in changes of substantial conditions, such as socio-occupational position, amount of obligatory activities or location of home and work. In the short term (which is the scope of our analysis), however, a number of substantial social, temporal and spatial conditions determine the range of choice of an individual in performing his daily path. Within the range of choice, an individual presumably tends to achieve that course of daily life which best fits the specific situation, say, in terms of saving travel time and other costs of linking the intended activities or in terms of enlarging the access to different optional activities.

Simplifying the generation of an individual's daily path, we arrive at the presentation of the process shown in Table 4. We distinguish between three process levels. At the highest level in the process hierarchy, each individual path is strongly affected by the collective socio-economic, temporal and spatial systems. The socio-economic system of activities and institutions (A) induces the differentiation between activities, division of labour and other tasks among the population groups, the structure of economic units in the given area and a mechanism which affects the individual's behaviour in production and consumption processes.

The temporal system of activities (B) embodies the general arrangement of different activity categories with regard to frequencies, collective schedules and episode duration. This arrangement is jointly shaped by social and biological clocks. The system of place categories (C) corresponds to the system of activities and institutions, defining at general level the differentiation between the urban spatial units with regard to their functions, distribution by size and specific usability. And finally, each individual daily path is very much influenced by the urban land use and transportation system (D) which, by the spatial patterns and technology of transportation connections, distributes the primary accessibility conditions among the social and spatial segments of the urban area.

At the lower levels of process hierarchy, under the influence of the urban systems and according to his own relative socio-economic position and household variables, each individual takes a relative position in space and time allocation. This implies the basic allocation of the individual's time as expressed in amount of occupational and other obligatory activities and their temporal fixities (G), as well as a basic allocation of the individual's activities to place categories (J). It also implies the individual's most recurrent geographical positions (F) which, in combination with his access to transportation means, define the relative ease of access to the remainder of the urban area (I).

While most of the individual's basic conditions for daily life can be steady for a certain period of time, the specific conditions for performing a daily path may vary from one day to the next. It may be shown that that this lowest process level, some path differences between comparable individuals are solely due to the frequency differences between the activity categories, resulting in various combinations of packages of activities and non-home sojourns occurring on the same day (M, N). At the single-day level, however, an individual presumably acts to improve the conditions of his daily life.

Thus, the individual's objectives in covering the path on a given day can also include some options in the time allocation (G) and the temporal order (K), through an appropriate locating and linking of activities in time and space (O).

PATTERNS EXHIBITED IN INDIVIDUAL DAILY PATHS

As pointed out before, the generation of the individual daily paths consists of the processes of urban life at various levels. Accordingly by observing the paths we face the whole set of interrelated patterns ranging from the highest level, for example, the division of labour and other tasks between population groups, to the lowest level, for example, the individual's options in connecting several activity places in one home-based loop on the given day.

In the following paragraphs we will present very briefly a few empirical examples (see the Appendix for a full list) from the explorative study carried out in Amsterdam, showing the variety of patterns in question and indicating the relevance of the path approach for the land use and transportation study.

Allocation of time. The allocation of time is certainly one of the most important issues of urban life processes. In any given stage of urban development, the differences in time budgets between the various population groups show in an implicit way the effects of both social and spatial differentations within an urban community. At the level of basic time allocation, to various main activity categories, the content of daily life is clearly related to the individual's social position (Table 5).

The time budgets of the main population groups differ greatly, primarily because of character of their daily tasks. As measured by an index, the time allocation differences between the main population groups are considerable, even when the time spent on physiological needs is included. The similarities of the time budgets of comparable groups from various samples are evident (Table 6).

Obviously, time allocation has to be studied in relation to other patterns, particulary the spatial ones. This permits us to show the effects of spatial conditions separately. In that context, information on existing time allocation can be used in modelling future development and estimating the potential changes in activity and travel demand. In the more detailed study, an important aspect is the portion of time spent on connecting different parts of the activity pattern.

Relationship between activity categories and place categories. In an integrated urban transportation study, insight in the relationship between time use and space use is of great significance for it shows travel need in the spatial organisation of activities. As shown in Table 7, the time spent on different activities is unevenly distributed among the place categories. Most evident is the polarisation between the home place, where little time is spent on work, and the non-home places, mainly used for the obligatory activities.

We also measured this association between activity categories and place categories in terms of the coefficient of contingency. The example in Table 8 shows the close association between time allocation and space allocation.

This pattern is obviously caused by the collective and institutionalised system of space allocation to activities. Most place categories in an urban area are highly specialised, the residential space being an exception for part of the activities. As Table 9 shows, the use of work places is rather one-sided, compared to the variety of time spent in homes.

The implications of these constraints in space use for the trip need are obvious. Generally, the more different activities an individual performs on a given day, the more place categories he needs. Consequently, a trip will occur more frequently. As illustrated in Table 10, the percentage of activity transitions associated with a trip is considerable. This aspect may also be of importance to the estimation of potential trip demand. For example, to meet the requirements of the possible growing variety of the individual's life in future, including many fine-meshed daily schedules, the relativity of trip demand should be analysed in relation to the relativity of space demand.

With an observed matrix of activity transitions and its distribution over place categories, the transitions occurring as trips can show the specific demand of various activity linkages, as illustrated in the example in Table 11. It is clear that the frequency of trip demand varies according to the kind of activity transition.

The majority of trip-demanding transitions are caused by the mutual separation of activities in non-home sojourns. Figure 1 shows the substantial difference between home sojourns and non-home sojourns with regard to sequencing the activities within the same address. As opposed to other sojourns, which in most cases are used for but one single activity, the home place facilitates the larger sequences of different activity episodes without a journey in between.

Temporal order of activities. Owing to the temporal arrangement of many activities, both by collective schedules and biological clock, the individual's choice in locating activities is greatly limited. The specific time periods of the day are assigned to many recurrent activity categories, so that the choice in using the remainder of time is firmly affected. We shall mention only two of the many aspects of the temporal order.

In Table 12 we use an example from our survey to show the meaning of the strong temporal order for the use of space. It is a well-known fact that the capacity of many urban functions is very much constrained by temporal concentration of activities. In fact, there is an association between the spatial segregation of activities into many single place categories and the assignment of specific day intervals to activity categories.

Another example shows the influence of the time order of daily tasks on the individual's choice in optional activities. As shown in Table 13, the frequency of non-home, non-obligatory activities of housewives is strongly affected by the length of the interval at their disposal.

The additional data on the same sample (omitted here) show that this relationship is close also in terms of action radius; the distances from home to the optional activities clearly increase with the increasing length of disposable intervals, thus showing the spatial aspect of choice. The implication is that the meaning of temporal constraints is similar to that of mobility/accessibility constraints, and this relationship deserves more attention in integrated planning for access to activities.

Compulsory and optional mobility. One of our basic assumptions in the study reported was that the various parts of an individual's daily path are interdependent. This can be seen in the frequency of different activities and their linkage through time allocation and other facets. One of these is the relationship between the individual's compulsory distance and the remainder of the distance travelled on the same day.

Given the travel budget of an individual, measured in terms of distance or time or generalised travel resources, the increase of distance to an obligatory activity may affect the remainder of the individual's travel on the same day. An empirical indication of this relationship is given in Table 14, showing how the substantial portion of the individual's mobility increase is being absorbed by the compulsory distances.

Obviously, this is shown up more pronouncedly in the less mobile groups. The absorption of mobility by the long distances to task activities is, of course, only an indicator of the more important consequences: the constraints of the individual's choice in access to other non-task activities.

Linkage behaviour in the individual daily paths. In the study of daily paths, special attention has been paid to the individual's behaviour with regard to linking a given number of non-home sojourns within the same day. With the same intented number of non-home sojourns per day the trip frequency can differ, owing to the differences in the portion of non-home sojourns reached from home. Actually, the number of trips for the (n) non-home sojourns can vary between (n+1) and (2n). While the number of ways in which an individual can link the non-home sojourns increases rapidly with their number ($C=2^{n+1}$), the distance relationship between various combinations of linkage (see Figure 2) seem to be an important factor in the individual's choice.

It can be shown that the higher the frequency of non-home sojourns, the longer the distance of the non-home sojourns from home, and the shorter their mutual distances, the larger home-based loops will be made. In fact, the individual's behaviour in making the temporal and spatial clusters of non-home activities reflects the joint effect of different parameters. Our survey has shown, for example, that many large individual non-home sequences are made by one or more individual transport means, and very few by public transport only.

It is clear that the meaning of basic concepts such as mobility and accessibility is modified when applied to the individual's linkage of activities. To express the joint task of transportation and land-use structure, we can also think of a new measure which will give us the degree of ease and range of choice in the individual's access to different activities.

If measured between all activities or place categories, the accessibility of urban structure is, actually, the degree of ease of the individual's transition between various activities, summed over the whole urban area and over an observed period. This quality - which we call transitiveness - can be measured, for example, as a number of different activities from which an individual can choose a specific transition within a given period, or as a distance or time required to complete an individual path connecting a specific set of activities in the urban area.

PRELIMINARY CONCLUSIONS AND RECOMMENDATIONS

Methods

□ Urban transportation research and planning should become an integral part of the urban social, temporal and spatial organisation. Such a position is essential in the generalised task of land use and transportation system which is the distribution of access to activities.

□ There is a need for a more direct approach by transportation research and planning to the daily life patterns, which are most evident at the individual's level. Both mobility and accessibility have to be analysed with reference to the variables of the activity system. The generation of travel need and trip pattern can be better interpreted when the temporal and spatial variation of the activity patterns is taken into account.

Objectives

□ Evaluation of the individual's choices deserves central position in the models of urban land use and transportation systems. These choices should be evaluated in terms of the ease of access to various (existing and potential) activities, as well as the range of spatial and temporal possiblities to perform various activity programmes.

□ Since daily travel for the regular and obligatory activities tends to absorb the major part of the mobility increase, the objectives of land-use and transportation planning should be primarily concerned with reducing the amount of travel for obligatory activities so as to stimulate the development of non-obligatory, non-home activities.

□ More attention should be paid to the fact that the individual's propensity to participate in the non-home, non-work activities is opposed/limited by the length of time interval at his disposal. Since the majority of disposable time intervals on a weekday are shorter than a couple of hours, the access to activity places should be adapted to these time constraints. More generally, however, the model could be developed to search for an optimal relationship between the distribution of disposable time intervals (affected by the collective time order) and the spatial distribution of access to activities.

Devices

□ More theoretical empirical work is needed to reduce travel by means of urban spatial restructuring, especially by mixing activities in space and time. The tendency to enlarge the urban scale and specialise place categories should be considered in terms of the implications for total community costs of access to activities.

□ One way of reducing the spatial and transportation over-capacities (owing to the concentration in time), as well as the distances between various urban functions, is to ease the time order of activities as much as possible.

□ Substituting communications for travel, work and study functions should be added systematically to the home place, which is the individual's central place in daily life and the most universal place category.

□ Since most time is tied to home and task address (work, school), the access to other (non-home, non-task) activities should be planned with regard to the joint/combined accessibility from home, task place and along the daily home-task place corridor.

□ In planning the intra-urban transportation system, models/technologies should be favoured which stimulate the proximity/merging of activities at fine scale. Also, preference should be given to the transportation system with the greatest linkageability, i.e. the ability to serve efficiently the connection of various places along an individual's daily path.

Table 1. Requirements for the scope of an urban transportation study

Subjects in the transportation study	Observing all trips (including walking trips)	Observing activities	Observing individual series
	Indispensable for the following parts of analysis/interpretation		
Trip frequency and purposes	Total trip production by activity and population group	Frequency of travel-demanding activities; differences in mobility and accessibility	Frequency of non-home arrivals as basic measure for trip demand
Choice of time and place of activity	Choice of place as affected by mobility	Relationship between disposable time interval and action radius; ratio travel time to sojourn time	Temporal and spatial clusters of destinations; distance behaviour in trip series
Travel costs v. choice of travel mode	Choice of travel mode as affected accessibility		Interdependence of trips with regard to mode choice; portion of travel time in total time budget

Table 2. The elements of an individual's daily path

Characteristics	The elements		
	Single activity episode (e)	Sojourn (S)	Trip (T)
Functional	Activity category (a)	Land use (u)	Travel mode (m)
Temporal	----------------------------------Time interval (t)----------------------		
Spatial		Location (p)	Geographical relation (r)

Table 3. The basic variables in the study of the individual's path

Personal and household variables
Sex
Age
Occupation
Educational level
Category of job or school
Number of persons in household, by age
Household income class
Number of vehicles in household, by category
Home location (square 500 x 500 m)
Job/school location
Home-centre distance
Home-job/school distance

Activity variables
Activity category
Activity place category
Location of activity place
Distance between activity place and home
Start time of a single activity episode
Duration of an episode
Start time of a sojourn
Category of the longest activity within a sojourn
Duration of sojourn

Trip variables
Travel mode
Start time of a trip
Duration of trip
Geographical relation (zone to zone)
Trip distance
Mean travel speed
Link number (order within a home-based loop)
Trip number (order within the individual's path)

Path variables
Number of single activity episodes
Length of a disposable interval, by time of the day
Number of trips
Number of sojourns by category
Number of home-based loops (chains) by size
Total potential distance by combination of linkage
Total travel time per home-based loop
Total travel time in a path
Action radius (distance from home to farthest sojourn)
Total mileage in a path
Number of travel modes used in a home-based loop
Position of trip with regard to home/task address

Table 4. Generation of an individual's daily path: a simplified presentation

A	Socioeconomic system of activities and institutions	
B	Temporal system of activities (frequences , duration , schedules)	
C	System of place categories	
D	Land use and transportation system	
E	Individual's socio-occupational and household variables	
F	Individual's home and task locations and home-task corridor	
G	Individual's basic time allocation and basic temporal order	
H	Individual's access to transportation means	
I	Relative locations of different place categories against individual's locations F	
J	Individual's basic allocation of activity categories to place categories	
K	Disposable time intervals of the individual	
L	Portion of the individual's travel budget absorbed by reaching task locations	
M	Individual's package of activities for the given day	
N	Individual's package of non-home sojourns for the given day	
O	Individual's choice between possible time / space configurations of daily package	

Table 5. Time allocation to different activities by two population groups: New West Amsterdam, random sample

Activity category	Workers	Housewives
	Percentages of 24 hours (weekday)	
Sleep, meals, personal care	40.1	43.0
Work, occupational activities	28.6	0.2
Housekeeping, errands	2.8	23.1
Shopping, services	0.6	2.2
Resting, conversation with family	4.9	7.8
Watching TV	5.9	7.6
Reading (book, magazine, paper)	3.5	3.9
Study, attending lectures	1.5	0.3
Taking a walk	0.6	0.5
Hobbies	1.6	2.4
Active sports	0.3	0.1
Other (leisure) activities	2.9	4.9
Travel (all modes)	6.7	4.0
Total	100.0	100.0

Table 6. Index of difference between the time budgets of various population groups +

		residents of New West Amsterdam			
		workers	housewives	students	pensioners
residents of new west amsterdam	workers		0.51	0.44	0.55
	housewives			0.57	0.51
	students				0.60
residents of old west amsterdam	workers	0.14			
	housewives		0.24		
	students			0.21	
	pensioners				0.32

+) Index of difference between the time budgets of groups (a) and (b) is calculated as :

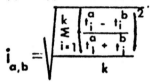

$$i_{a,b} = \sqrt{\frac{\sum\limits_{i=1}^{k}\left(\dfrac{t_i^a - t_i^b}{t_i^a + t_i^b}\right)^2}{k}}$$

wherein :

t_i^a = time spent by group (a) to activity (i)

k = number of activity categories

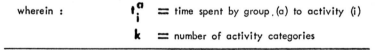

Table 7. Distribution of time spent in selected activities among different place categories. New West Amsterdam, random sample, population aged 15 to 80

Activity category	Place category						
	Home	Another dwelling	Job address	School	Shops and services	Other	Total
	Percentage of time per activity category						
Meals	84	1	8	1	1	5	100
Personal care	97	0	1	0	1	1	100
Work	4	1	84	0	2	9	100
Study	35	2	0	53	0	10	100
Housekeeping	97	2	0	0	1	0	100
TV	98	1	0	0	0	1	100
Reading	96	2	1	0	0	1	100
Hobbies	88	4	1	0	0	7	100

Table 8. Relationship between activity categories and place categories as measured by the coefficient of contingency in time spent*

Population group	Sample area	
	Old South	New West
	Amsterdam	
Workers	0.884	0.886
Housewives	0.894	0.894
All groups	0.894	0.899

* Based on a contingency table with 30 activity categories and 30 place categories

Table 9. Five longest activities at three place categories: New West Amsterdam, workers, random sample

Place category	Activity category	Percentage of total time spent at that place category
HOME	Sleep	56.5
	Watch TV	10.0
	Meals	6.4
	Reading	6.0
	Personal care	4.2
	Total five longest	83.1
ANOTHER DWELLING	Visit	52.4
	Game	12.5
	Work	7.4
	Errands	5.9
	Watch TV	4.2
	Total five longest	82.4
WORK PLACE	Work	93.5
	Meals	3.8
	Rest	1.8
	Reading	0.2
	Personal care	0.1
	Total five longest	99.4

Table 10. Mean frequency of activity transitions
and portion of transitions associated with a trip:
Two districts of Amsterdam; random sample

Population group and sample area (district of Amsterdam)	Mean number of activity transitions per person-day	Percentage of transitions associated with a trip
Workers, Old South	14.6	38.4
Workers, New West	13.4	39.6
Housewives, Old South	18.5	27.0
Housewives, New West	17.5	28.6
Pensioners, Old South	15.3	23.5
Pensioners, New West	14.4	26.4

Table 11. Trip demand of different activity transitions:*
Old South Amsterdam, random sample, population aged 15 to 80

From To	Work	Study	House	Errand	Meal	Rest	Social	TV	Read	Hobby	All†
	Percentages activity transitions associated with a trip										
Work		...	...	78.7	35.4	42.7	83.4	75.0	88.5	...	63.6
Study	...		...	93.3	51.2	35.5	75.0	26.9	28.6	...	51.9
Housekeeping	78.2	22.7		35.2	0.1	2.1	49.5	0.1	2.1	6.2	17.8
Errands	92.5	92.8	36.6		50.6	67.4	77.2	58.5	46.7	...	73.0
Meals	52.5	36.8	1.0	71.2		6.6	60.0	2.7	3.5	20.9	21.8
Resting** (with family)	20.8	26.4	2.8	58.0	5.4		60.6	3.5	0.0	3.8	21.7
Social** (with friends)	...	...	50.0	92.8	46.7	53.4		60.6	39.4	20.0	66.2
TV	89.5	19.2	1.7	33.3	3.8	6.8	42.1		2.9	9.1	16.9
Reading	53.0	38.7	2.6	37.1	1.2	5.3	41.2	0.0		7.4	17.8
Hobbies	...	...	12.9	61.5	2.0	10.9	61.5	6.1	11.1		22.9
All activity categories	63.5	48.5	22.4	72.5	16.2	27.4	65.4	11.8	18.0	20.8	37.3

*   This table is derived from 610 kinds of activity transitions observed in a 38 x 38 matrix

†   Including the transitions not reported in this table

** Social activities with friends and relatives

Table 12. Temporal concentration of use of selected place categories:
New West Amsterdam, random sample, population aged 15 to 80

Place category	Mean percentage of persons using a place between 6 a.m. and 12 p.m.	Maximum percentage observed between 6 a.m. and 12 p.m.	Ratio maximum to to mean percentage
Schools	3.6	10.2	2.8
Shops elementary	1.3	4.4	3.4
Other shops	0.7	2.4	3.4
Medical services	0.7	2.4	3.4
Leisure/recreation	1.7	5.9	3.5
Traffic space	6.6	16.3	2.5

Table 13. Relationship between the length of disposable interval* and the percentage of
intervals spent away from home: Housewives, three districts of Amsterdam

The length of disposable interval (hours)	Time of the day	
	Before 6 p.m.	After 6 p.m.
	Percentage of intervals partly or entirely spent away from home	
≦ 1	2.6	0.8
1 - 2	25.7	4.6
2 - 3	41.0	5.9
3 - 4	76.0	15.0
4 - 5	83.0	18.8
> 5	100.0	33.3

* A disposable interval is defined here as a time difference between the end of an
obligatory activity and the start of the next one

Table 14. Portion of total mileage spent on action radius:
Employed men, three districts of Amsterdam

Action radius* (km)	Car owners mean ratio to total mileage	Non-car owners (2 x action radius) a person a day
≦ 4	0.65	0.68
4 - 6	0.69	0.81
> 6	0.73	0.93

* The term "action radius" denotes the distance between the individual's home and the
farthest place visited on the given day (in most cases a work place), subject to his
first departure and last arrival on that day being at the home place

REFERENCES

[1] The research reported here was initiated with the Department of Public Works in
Amsterdam in 1968. It is at present known as the research project SATURA (Space and Time
Use Research Amsterdam). In the meantime, samples from five districts of Amsterdam have
been observed. Closely related to this project is the Analysis of the Individual Trip
Series, active since 1972 and developed with the University of Technology in Delft.
[2] For the purpose of this paper, which is only a very brief presentation of the research
in question, we do not systematically refer to a number of different approaches developed
elsewhere. Most other work of similar scope has been taken beyond the field of
transportation discipline. Specifically, the disaggregate time-space approach was
initiated by Hagerstrand (1970, 1973). See also Hanson (1977), Jones (1976), Lenntorp
(1976), Matzner (1976), Parkes (1975), Thrift (1977), and Westelius (1972). Also related
to our subject is the work on the partial models of trip linkage developed by Gilbert
(1972), Marble (1964), Nystuen (1967), Sasaki (1972) and Vidakovic (1974, 1977).

Gilbert, G., Peterson, G.L., and Schofer, J.L., "Markov renewal model of linked trip
travel behaviour" (1972), in Transportation Engineering Journal.

Haggerstrand, T., What about people in regional science? (1970), Papers of the
Regional Science Association, Vol XXIV.

Haggerstrand, T., The impact of transport on the quality of life (1973), report on the International Symposium on the Theory and Practice of Transport Economics, Athens.

Hanson, S., Urban travel linkages: a review (1977), a resource paper prepared for the Third International Conference on Behavioural Travel Modelling, Tanuda, Australia.

Jones, P.M., The analysis and modelling of multi-trip and multi-purpose journeys (1976), Working Paper 6, Transport Studies Unit, University of Oxford.

Lenntorp, B., Paths in time-space environments: a time geographic study of movement possibilities of nidividuals (1976), Lund Studies in Geography.

Marble, D.F., A simple markovian model of trip structures in a metropolitan region (1964), Papers, Regional Science Association, Western Section.

Matzner, E., Henseler, P., and Rusch, G., "Theoretical frame of references to the planning of urban commuter traffic" (1976), Egon, Matzner, Gerhard Rusch (eds) Transport as an instrument for allocating space and time - a social science approach, Technical University of Vienna.

Nystuen, J.D., "A theory and simulation of intraurban travel" (1967), W.L. Garrison and D.F. Marble (eds) Quantitative Geography, Part 1, Evanston.

Parkes, D.N., and Thrift, N., "Timing space and spacing time" (1975) Environment and Planning, Vol 7.

Saski, T., "Estimation of person trip patterns through Markov chains" (1972), Newell, G.F. (ed), Traffic flow and transportation.

Thrift, N., "An introduction to time geography, concepts and techniques" (1977), Modern Geography No. 13.

Vidakovic, V.S.," A harmonic series model of the trip chains". (1974), Buckley, D.J. (ed), Transportation and Traffic Theory, New South Wales University.

Vidakovic, V.S., A distance parameter of the trip-chain process (1977), Sasaki T. and Yamayoka (eds), Proceedings of the Seventh International Symposium on Transportation and Traffic Theory, Kyoto, Institute of Systems Science Research.

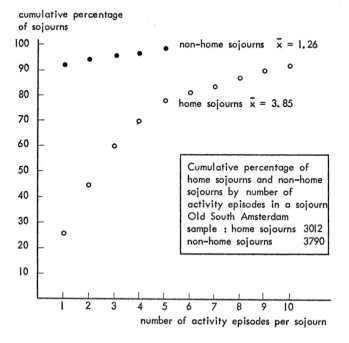

Figure 1

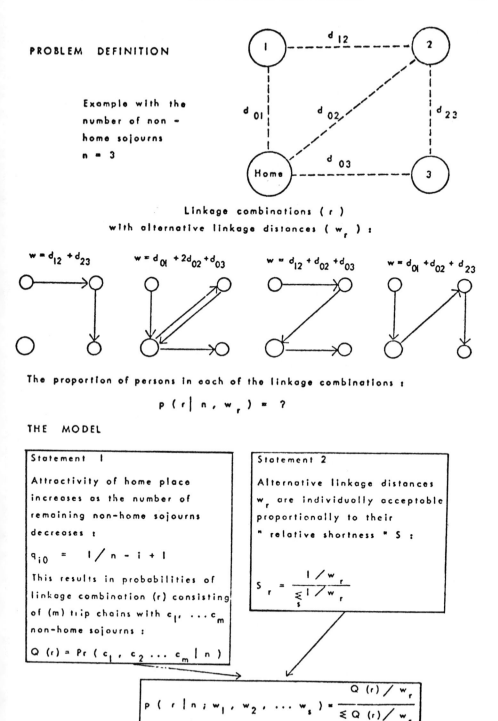

Figure 2. The individual's choice between alternative linkage combinations; partial model of frequency and distance behaviour in daily paths

APPENDIX: PATTERNS ANALYSED IN THE STUDY OF THE INDIVIDUAL'S PATH

□ Allocation of time to various activities and place categories related to personal and household variables.

□ Participation in various activities related to personal and household variables.

□ Frequency distribution of the episodes by duration and activity category.

□ Frequency distribution of sojourns by duration and sojourn category.

□ Relationship between activity categories and place categories.

□ Activity transition matrix and portion of trips by transition category.

□ Relationship between activity sojourn category and time of the day; the strength of time order related to personal and household variables.

□ Frequency of non-home sojourns by category, related to personal and household variables and "daily package" of sojourns.

□ Number of different sojourn categories in a path and their relative locations.

□ Trip frequency and distance behaviour in relation to the configuration of non-home sojourns.

□ Modal choice in trip linkages.

□ Frequency and locating of non-home, non-task activities, related to home and task locations.

□ Hierarchy of distances in an individual's path; relation between action radius and total mobility.

□ Ratio of travel time to non-home sojourn time.

□ Allocation of time to sojourns at various distances from home.

□ Activity pattern and distance behaviour in using disposable time intervals of different length.

Some problems in forecasting car ownership for urban areas

15

K. J. Button and A. D. Pearman   United Kingdom

SYNOPSIS

This paper presents findings in a large on-going research project on car ownership being undertaken at the Universities of Loughborough and Leeds in the United Kingdom under the auspices of the Social Science Research Council. The paper is an extension of earlier work by the authors and offers new findings on the relationship between urban change and car ownership levels and patterns. Initially, the limitations of current "best-practice" methods of car ownership forecasting, as employed by traffic engineers in large-scale urban transport planning exercises in the United Kingdom, are discussed. The assessment considers not only the theoretical limitations of these techniques, but also their usefulness as actual tools for forecasting. In the past there has been considerable academic debate on several inherent internal inconsistencies in these models, but far less discussion of their practical usefulness in actual forecasting. Particular emphasis is placed on the influences exerted by different land-use patterns on vehicle ownership and the impact that changes in these patterns are likely to have on future ownership levels. It is argued that any form of development is likely to affect the general level of accessibility enjoyed by those living in the urban area and this, in turn, will influence car ownership decisions.

Empirical evidence, both for illustrative and calibrative purposes, of the problems associated with forecasting levels of urban car ownership is drawn from a number of sources. Close association with WYTCONSULT, however, has permitted the study team to draw extensively on data collected as part of a large land-use/transportation study carried out in the West Riding of Yorkshire (England). The latter data source provides useful and detailed information at the local level and permits a number of specific themes to be explored in some depth.

A series of models has been calibrated using logit type regression analysis. It is clear, from a technical point of view, first, that there is a need for some prior stratification of the population if the models are to be statistically sound and their parameters reliable; and, secondly, when logit analysis is used, some rather arbitrary assumptions must be made to permit calibration at all. The quantitative results based on the West Yorkshire data indicate specifically the influence exerted by accessibility, as measured by several objective indices, on car ownership levels in the region.

CAR OWNERSHIP MODELLING

Car ownership has long been recognised as an important determinant of overall travel demand. For this reason, car ownership variables play a central role in the traffic forecasting models employed by traffic engineers in urban transport planning activities. In particular, car ownership forecasts are important inputs of both the trip and model split sub-models of the standard sequential forecasting exercise. Further, recent advances in travel-demand modelling, which are both more disaggregated and explicit in their approach than the traditional four-stage recursive procedure, require considerably more detailed prediction of future levels of household vehicle ownership than have been needed in the past.

The realisation that car ownership has such a central position in travel-demand forecasting has led to increasing efforts to improve vehicle ownership models. Two broad schools of forecasting have grown up in recent years. The respective characteristics of these approaches can be illustrated by drawing on the experience of the United Kingdom. Until recently the standard method of forecasting, both at the national and local level, was the logistic curve procedure developed by the Transport and Road Research Laboratory (TRLL).[1] The technique itself was initially a simple extrapolation procedure which assumed a long-term logistic growth in car ownership per capita towards some eventual and exogenously determined saturation level. However, the exclusive reliance on a time trend, combined with debate on both the appropriate functional form of the sigmoid growth path and the eventual saturation level, led to some subsequent changes in the procedure. Initially, tentative steps were taken both to incorporate economic variables in the model (specifically income and vehicle price) and, secondly, to obtain more stable estimates of the saturation level of per capita vehicle ownership.[2] Later, further modifications were made on both these counts (e.g. the vehicle price variable was altered to incorporate allowance for usage costs) but, in addition, the functional form was changed from a logistic to a power growth curve.[3]

While this type of extrapolation approach may be of use at the national level, its adoption for urban planning exercises seems inappropriate. In particular, the procedure allows only very limited account to be taken of the impact of any changes in the transport system which may be initiated by the transport engineer. Further, it is designed specifically for national predictions and can only be calibrated from a long-time series of data. Urban traffic engineers have no such data available and must, at best, rely on a "snapshot" of the existing situation taken from a single transportation survey. While, to a limited extent, the first problem may be circumvented by modifying the usage-cost variable in the extrapolation procedure, the second difficulty is insurmountable.

Since time series models appear impractical in the context of urban traffic forecasting, cross-sectional techniques have begun to be employed more fully. In the United Kingdom this trend has received strong support through criticisms of the extrapolation forecasting used in inter-urban road investment appraisal (which has recently been fully explored by the Leitch Committee[4]). For example, the recently constructed Regional Highway Traffic Model (RHTM), although intended for aggregate rather than local urban forecasting, has a car ownership forecasting sub-model which has been designed to relate car ownership directly to a set of causal variables.[5] While the modified TRRL approach feeds in exogenously estimated income and price elasticities, the RHTM calculates such parameters as part of its calibration process. Consequently, the RHTM, if adopted at the urban level, would permit local, rather than national, parameters to be employed in the subsequent forecasting. In detail, the RHTM is a log-logit model relating the log of the logit of a household not owning a car

$$\log \left(\frac{P_0}{1-P_0} \right)$$

to a set of independent variables expressed in log form. Calibration of the RHTM car ownership model is by maximum likelihood procedures.

The RHTM car ownership forecasting framework has the clear advantage that it is, in broad terms, a causal model and can be specified in such a way that the output produced is suitable for modern disaggregate traffic forecasting models. Further, it should be possible to incorporate, with the set of explanatory variables, variables which are sensitive to, and ipso facto reflect, measures introduced by traffic engineers. (The appropriate form of such variables is discussed below.) It should not be forgotten, however, that there are costs in adopting a causal approach to car ownership modelling. On the technical side, for example, severe computational problems arise when it is necessary to group data prior to calibration. Further, the open-endedness of cells following grouping can lead to a degree of arbitrariness in the estimation procedure. On the forecasting side, the basit TRRL method simply requires knowledge of future population levels to provide predictions of the aggregate level of car ownership. By contract, causal models require prior knowledge of future values for each of the explanatory variables which may themselves require forecasting procedures as complicated as the car modelling exercise. In terms of cost-effectiveness, the TRRL method is cheap and relies essentially on national parameters calibrated from readily available data and then employed at local levels, but the RHTM approach demands local survey work to permit unique models to be calibrated for each urban area. It should be borne in mind, however, that modelling costs are normally dwarfed by the enormous costs of present-day urban transportation engineering projects.

POLICY-SENSITIVE VARIABLES

The standard variables included in local car ownership forecasting models usually relate to household characteristics and especially to the level of income of the household. In fact, income has been included in a number of different forms in attempts to allow both for the supposed luxury nature of vehicle ownership[6] and the high initial capital cost of ownership.[7] Such household variable are, however, outside the control of the traffic engineer and, although it is clearly important to make correct allowances for them when forecasting, they are, therefore, of secondary importance when assessing the impact of various traffic control options, or when ranking investment priorities.

To date, attempts to incorporate policy-sensitive variables in car ownership models have been rather incomplete. A generalised cost-of-car ownership and usage variable of the type used in the recently modified TRRL approach (see above) is of little use at the urban level, for it reflects general temporal trends rather than specific spatial differences in transport costs. Urban models have tended to rely on proxy variables to allow for policy effects and, in particular, widespread use has been made of spatial variables. One such proxy is population density which, it is argued, is a factor that land-use planners can control. Population density has variously been thought to reflect adequacy of local facilities (and hence to be negatively related to travel need), quality of public transport, or the level of congestion on highways. Distance from the urban centre is sometimes adopted for similar reasons.

The limitations of such surrogates are clear. They do not directly reflect the influence of traffic engineering measures and so their value in forecasting models is at best doubtful. Further, the exact causal links involved are unclear. Indeed, car ownership is itself frequently used by regional scientists when emphasising factors such as population density and residential location.

More explicit instructions of policy-orientated variables have been attempted, usually with accessibility indices of some type. Dumphy in America, for example, found a strong correlation between car ownership and a crude public transport accessibility index based on the percentage of jobs reached from an area in forty-five minutes.[8] Shepherd used data derived from a series of Australian cities to calibrate simultaneous equation models of urban travel where car ownership was treated as a function of public transport accessibility defined in terms of route miles per 1000 of the population and service miles per route mile.[9] Fairhurst has related household car ownership to a frequency index of public transport in the Greater London area of the United Kingdom.[10]

The general approach described in the previous paragraph is taken as a basis for the work reported here. The empirical results given in the following sections are based on two major assumptions. First, that household-based causal models provide the appropriate framework within which to model car ownership. Secondly, that, since car ownership or use is frequently a target for traffic engineering and is, in any case, almost certain to be influenced by it, it is essential that policy-sensitive variables be explicitly incorporated in the model. Further, there is a need for these variables to be readily interpretable, so that traffic engineers can assess the potential impacts of various policy alternatives.

THE WEST YORKSHIRE STUDY

The West Yorkshire Transportation Study provides a very good data base on which to build useful car ownership models.[11] Not only has it been possible to use the data collected from nearly 10,000 households to construct causal models of vehicle ownership, but the survey also provides sufficient information about the transport system to enable the incorporation of policy-sensitive variables in the analysis. The household data were collected between 22 April and 18 July 1975 from some 12,322 addresses in the West Yorkshire area. Of the sample, some 9963 households were interviewed and, of these, 7812 gave sufficient information to permit detailed analysis to be carried out (specifically, all these households gave information about their income). One noteworthy feature of the sample is its high proportion of pensioner households (about 22 per cent of the total), which gives a bi-modal income distribution.

The analysis itself follows two broad lines. Initially, cross-classification (or category analysis) techniques are used to provide some preliminary breakdown of the data. Secondly, log-logit analysis is applied to look at the causal relationships in more detail and, in particular, to examine the influence that different policy-sensitive variables, in combination with expected income changes, can exert on the vehicle ownership level of an area.

THE VARIABLES

The basic variables in the study can be divided into two broad groups: those that relate to the socio-economic characteristics of the households under consideration, and those - the policy-sensitive variables - which may in some way be acted on by transportation engineers or urban planners. Since it is important in urban transport planning to have forecasts not only of average household car ownership but also the number of no-car, one-car and multi-car households, a number of different dependent variables were defined. Further, since availability and not strictly ownership influences travel patterns, this was selected as the appropriate measure for analysis. Therefore, the attributes of car ownership/availability chosen were: C = the average number of cars or vans available per thousand households; P_0 = the percentage of households with no cars/vans available; P_1 = the percentage of households with one car/van available; and P_2 = the percentage of households with two or more cars/vans available.

Variables reflecting the socio-economic attributes of a household were taken to be the following:

□ Y = the household income level measured in £ sterling, divided into eight groups, six with widths of £1040, then £6241-7800, and finally a residual class of over £7800

□ E = the number of employed residents in the household

□ H = the number of household residents

□ E/H = the ratio of employed to total household residents

□ S_1 = a household structure breakdown (code 1) where we have:

1 resident, all pensioners, 0 or 1 employed

2 residents, all pensioners, 2+ employed

3 residents, not all pensioners, children present, 0 or 1 employed

4 residents, not all pensioners, children present, 2+ employed

5 residents, not all pensioners, no children present, 0 or 1 employed

6 residents, not all pensioners, no children present, 2 + employed

□ S_2 = a household structure breakown (code 2) where we have:

1 : 0 employed residents and 1 non-employed resident

2 : 0 employed residents and 2+ non-employed residents

3 : 1 employed resident and 0 or 1 non-employed resident

4 : 1 employed resident and 2+ non-employed residents

5 : 2+ employed residents

Four policy-sensitive variables were examined. Two of these are of the traditional spatial type, while the others are of a more direct kind and relate more closely to the type of variable which can be influenced by the transport or traffic engineer. The variables are:

□ Z = the type of zone in which a household is located

1 : urban/suburban areas

2 : dormitary/rural areas

3 : other areas

□ D = the household residence zone population density, split into four roughly equal groups by increasing density

□ T = the subjective estimate of journey to work time by public transport (including walk) for the head of the household (irrespective of the actual mode used). The variable is classified as follows:

1 : 5 minutes or less

2 : 6-10 minutes

3 : 11-15 minutes

4 : 16-30 minutes

5 : 31-45 minutes

6 : 46-60 minutes

7 : more than one hour

☐ G = the mean reduction in generalised cost to households in that residence zone arising from having more than 0.6 cars available per driving licence, compared to being dependent on public transport, assuming a typical distribution of journeys to work. This measure splits households into four equal groups:

1 : those who gain up to 18.2 generalised cost-minutes from car availability

2 : those who would gain 18.2 to 20.7 generalised cost-minutes from car availability

3 : those who would gain 20.7 to 27.4 generalised cost-minutes from car availability

4 : those who would gain more than 27.4 generalised cost-minutes from car availability

Before we look explicitly at the importance of the policy-sensitive variables, a number of preliminary comments are necessary on the importance of the socio-economic attributes of households. Tables 1 and 2 give breakdowns of the car availability of households in terms of several of these variables. To overcome the worst problems of sampling variations, only cells with 10 or more observations are recorded in the tables. Nevertheless, the standard errors of cells still remain relatively large. It is apparent from the tables that, as one would expect, there is a positive relationship between car availability and household income, and that this relationship remains fairly firm even when allowance is made for differences in other socio-economic variables.

Comparisons of the various non-income attributes may be made. The household structure index, S_2, shows a definite tendency for those households coded highest to have car availability, although the effect is much less consistent, as we see in the body of Table 1, since allowance is made for income variation. The use of employed residents in Table 2 produces patterns similar to those of S_2 when income is controlled. This suggests complicated interactive effects with Y. This would seem to indicate that there are dangers in using these types of variables together in regression-type models of car ownership, certainly if they are calibrated at the household level.

The introduction of H in Table 2 produces results similar to those of E when used separately from it, although the order of magnitude of the observed effect is considerably less. The combination of E and H as classifying variables produces some interesting additional results, however. The "all incomes" column shows that for a given H, C increases with E. This, however, is largely a consequence of increased employment which provides households with a higher income (when H and Y are held constant, C more often than not falls as E increases).

In view of the above findings, it was decided to investigate in more detail the use of various measures of household structure to classify households in later analysis. The three classification variables examined were E/H, S_1 and S_2. E/H has the disadvantages both of taking the value zero when E = O and of giving the same value for households with H = 2 and E = 1 as for households with H = 4 and E = 2. The classification S_1 gives less variation than S_2 because even its lowest code permits one employed resident, and the presence of children makes little difference (e.g. between categories 3 and 5). For these reasons, classification S_2 seems the most useful indicator of household type and is therefore employed in the majority of the subsequent calculations.

A problem that becomes apparent when this type of breakdown is attempted is that there is a marked difference in the behaviour of small and poor households, compared to the majority of the results derived. The general implication of this is that it cannot simply be assumed that future increases in income will result in these groups exhibiting car ownership patterns similar to those of households currently enjoying much higher levels of income. It seems that such households are in many ways atypical and that somewhat different types of models may be required if reliable forecasts of their future vehicle acquisition are to be obtained. This is certainly an area requiring much more detailed research than it has attracted in the past.

THE IMPORTANCE OF POLICY-SENSITIVE VARIABLES

Prior to the calibration of any models, further cross-classification analysis was carried out to examine the usefulness of the different policy-sensitive variables. Table 3 shows the impact of zonal type when both income and the numbers of employed residents are controlled. Although this variable has been explored in some depth,[12] it is not considered suitable for disaggregate car ownership forecasting purposes. As can be seen in Table 3, the influence exerted by the different variables is blurred. Secondly, and possibly of greater importance, while the urban planner may have long-term control over the general development of land-use policy, such controls are not really sensitive enough to be realistically considered as "policy instruments" to regulate local traffic.

Population density, the second policy-sensitive variable, was derived by aggregating the 1816 fine zones of the WYTCONSULT study to form 300 new zones for which areas had been estimated from maps, and then taking the ratio of census population figures for these areas. The results were grouped into four roughly equal categories ranked by increasing density. In Table 4 we see that C and D are negatively related, regardless of Y, except that density seems to have less effect for lower incomes. P_0 and D seem to be positively related, except again for the very lowest income group. Again, however, we must question the usefulness of such a crude tool as a "policy variable".

Variable G was also calculated on the basis of about 300 aggregated zones for which we have information available on the generalised costs of journeys to five different types of work by four different levels of car availability. The generalised cost of the 0.6 + car/licence category was taken from that of the no-car available category and weighted by the average frequency of the five job types in West Yorkshire. This was coded into four roughly equal bands, with code numbers increasing with the gain from car ownership. As expected, we see in Table 4 that C is positively and P_0 negatively related to G. Overall, the G variable does not, however, appear sufficiently sensitive to provide a useful tool for policy analysis. In particular, there are negligible differences between bands 1 and 2 of the variables.

Variable T is based on a smaller sample than either G or D, because only half the heads of households were questioned on the topic and half of these did not make work trips. The remainder were asked to estimate their journey to work times by bus and walk. These times were available in seven bands (as recorded above), unequal in length of time and frequency of observations. Careful examination of the frequencies confirms the first impression that the variability of car ownership in relation to T is greater than in relation to either D or G. As expected, the relationship between C and T is positive, and that between P_0 and T negative. For journeys up to fifteen minutes (i.e. T bands 1 to 3), car availability is not much affected by increased journey to work times and there is a suggestion that for those within five minutes of work, car ownership is somewhat higher than for those with slightly further to go. The limited data available on this variable, however, prevent its use in the later analysis and G was, therefore, used in the more detailed modelling which now follows.

While the category analysis provides useful insights into the relationship between car ownership and various policy variables, it is rather a restricted form of modelling framework.[13] Logit analysis offers a much more useful method of handling household vehicle ownership decisions, especially since non-linearity in the relationship between car ownership and income has been detected.[14] A quasi-logit model of the form

$$P_0 = \frac{1}{1 + e^{-(b_0 + b_1 \log Y)}}$$

was used which, for calibration purposes, was transformed to become:

$$\log \left(\frac{P_0}{1 - P_0} \right) = b_0 + b_1 \log_{10} Y$$

The equation was calibrated over the fourteen income groups for each combination of our five household types (S_2) and four accessibility bands (G). Some statistical problems arise from doing this. First, since we are dealing with grouped data it is important to ensure that in-group variation is small compared to between-group variation. This is achieved by grouping the data together in income bands rather than in geographical zones, as is usual in transportation applications. Secondly, for some income cells the frequency of no-car ownership was zero, which prevented estimates of the log-logit being obtained. To correct for this, a slightly modified equation form was employed:[15]

$$\log \left(\frac{P_0 + \frac{1}{2} n}{1 - P_0 + \frac{1}{2}n} \right) = b_0 + b_1 \log Y$$

Here, n is the number of households in that income/household type/accessibility band cell. Thirdly, the homoscedastic assumption of regression analysis may be threatened by the grouping exercise. Therefore, to allow for the different numbers of observations in each cell, a weighting scheme was adopted to weight each group by the inverse of its variance. Finally, information is available on the midpoint of each income group but not for the actual mean income. This is likely to bias the regression coefficient estimates and, in particular, to understate their absolute value. To allow for this, the regressions were performed with estimates of the means derived from the overall income distribution of the sample.

A specific problem associated with the computer program employed (Statistical Package for the Social Sciences, Version 5) is that it assumes that the weights adopted imply that the number of observations have been replicated at exactly that point, rather than that the point is a mean with a variance around (which the weight is attempting to equalise for all points). Hence, the error variance will be grossly underestimated and R^2, t and F statistics misleading. As we see in the results produced in Table 5, however, inspection suggests that the estimates of b_0 and b_1 vary widely, but in an inderdependent way, so that their ratio, which allows us to determine the equi-probability income (at which $P_0 = \frac{1}{2}$), is less variable.

In Table 5 we see that, overall, the equi-probable income is negatively related to the accessibility index G, which means that the greater the generalised cost gain on work trips from car availability, the lower the level of income at which a household is just as likely to own a car as not. The results by household type are less clear. For household type 1, for example, the small number of degrees of freedom with few observations around the $P_0 = \frac{1}{2}$ level produce very erratic results. If we ignore this group, however, there seems to be a tendency for the equi-probability income to be lower for large households.

We see that for "all households" the estimates of b_0 and b_1 appear to increase in absolute value as accessibility gains (from car availability) increase. This seems to imply that the income elasticity for car ownership increases as more opportunities for generalised cost gains from car usage become available. Overall, however, it is clear that detailed analysis of the data in this way will only be reliable once the parameters of the quasi-logit model have been estimated in a more rigorous way, i.e. using maximum likelihood rather than regression methods.

CONCLUSIONS
This paper has attemped to illustrate a number of points. First, that car ownership levels are not simply a function of time but depend on a number of variables. Among these variables are several which may be influenced by urban planners and traffic engineers. Consequently, any policy to modify local transport systems will both influence the level of car ownership and also be influenced by it. The empirical work tests several type of policy-sensitive variables and provide evidence of their influence. Further, it is suggested that some of the cruder accessibility measures incorporated in certain car ownership models may be descriptive, but really do not provide many useful insights for those actually responsible for urban transport planning.

Further, it is apparent from the analysis of the West Yorkshire data that significantly different patterns of behaviour are exhibited by small and poor households, compared to the population as a whole. This would suggest that, possibly, no single forecasting framework can be derived for the entire population but that certain groups within it require special treatment.

Finally, the log-logit type of approach offers a very useful and flexible technique for modelling car ownership on the basis of local sample survey data. Associated with it are a number of practical problems which require some quite important modifications to the basic approach if a useful forecasting tool is to be developed. In several respects the corrections required are of a subjective rather than mechanical nature, which suggests that the techniques should be employed with a degree of circumspection.

Table 1. Average number of cars per 1000 households

S_2 Household structure Code 2	Annual household income, Y								All incomes
	Under £1041	£1041 - £2080	£2081 - £3120	£3121 - £4160	£4161 - £5200	£5201 - £6240	£6241 - £7800	Over £7800	
1	35	132	385	-	-	-	-	-	50
2	113	229	741	833	-	-	-	-	222
3	205	276	587	804	889	1154	1364	1765	449
4	357	429	684	986	1181	1456	1579	2091	765
5	636	442	629	794	956	1231	1358	1661	828
All	72	312	638	829	992	1263	1372	1737	558

Table 2. Average number of cars per 1000 households

H	E					Y				
		Under £1041	£1041 - £2080	£2081 - £3120	£3121 - £4160	£4161 - £5200	£5201 - £6240	£6241 - £7800	Over £7800	All incomes
1	All	43	176	556	630	-	-	-	-	133
	0	33	140	417	-	-	-	-	-	50
	1	197	192	571	692	-	-	-	-	353
2	All	144	284	597	832	908	1219	1317	1519	503
	0	125	220	647	-	-	-	-	-	209
	1	220	328	595	897	917	-	1400	1750	512
	2	-	391	592	813	906	1215	1333	1286	762
3	All	211	476	658	845	986	1424	1382	1679	771
	0	91	341	-	-	-	-	-	-	282
	1	273	439	661	1000	1160	1286	-	-	698
	2	-	566	660	845	958	1667	1312	1562	852
	3	-	923	621	639	936	1286	1308	2000	943
4+	All	100	380	675	823	1034	1230	1371	1843	809
	0	0	172	-	-	-	-	-	-	253
	1	-	429	700	952	1192	1500	1500	2235	805
	2	-	413	641	813	1053	1288	1480	1852	807
	3+	-	263	684	645	909	983	1293	1667	924

Table 3. Average number of cars per 1000 households

Z	Employed residents				Annual income of household					
		Under £1041	£1041 - £2080	£2081 - £3120	£3121 - £4160	£4161 - £5200	£5201 - £6240	£6241 - £7800	Over £7800	All incomes
1	0	52	162	696	-	-	-	-	-	109
	1	145	286	538	731	955	1160	-	-	455
	2	-	315	553	675	897	821	1218	1518	650
	3+	-	313	545	533	648	879	1103	1524	744
	All	66	249	551	674	835	928	1172	1489	430
2	0	157	351	-	-	-	-	-	-	274
	1		429	808	1185	1000	-	-	-	920
	2	-	867	694	929	1136	1700	-	-	1016
	All	145	531	800	1000	1400	1667	-	-	884
3	0	51	239	581	-	-	-	-	-	127
	1	354	370	697	988	1134	1486	1571	2160	696
	2	-	493	684	914	995	1475	1415	1647	884
	3+	-	333	700	712	989	1173	1487	1783	1012
	All	72	344	688	908	1025	1390	1471	1843	624
All	All	72	311	640	822	988	1262	1358	1754	557

Table 4. Average number of cars per 1000 households

Accessibility variable	Value of accessibility variable	Under £1041	£1041 - £2080	£2081 - £3120	£3121 - £4160	£4161 - £5200	£5201 - £6240	£6241 - £7800	Over £7800	All incomes
					Annual household income					
	1	95	419	700	933	1136	1490	1500	2000	692
	2	54	302	751	874	983	1368	1463	1707	620
	3	52	305	588	827	926	1167	1231	1625	527
	4	72	229	494	633	771	769	1138	1100	377
	1	75	294	590	765	902	1104	1214	1556	496
	2	48	288	586	716	928	906	1235	1828	481
	3	68	306	637	833	982	1476	1513	1656	565
	4	99	368	737	986	1130	1440	1529	1855	689
	1	83	236	561	720	1294	-	-	-	464
	2	650	177	373	605	762	-	-	-	451
	3	0	225	477	636	619	-	-	-	432
	4	176	222	484	705	900	1133	1154	-	538
	5	-	280	614	719	1224	1467	1400	-	724
	6	-	484	763	781	1167	875	-	-	776
	7	-	857	804	1026	1115	1292	1462	1833	1083

% of households with no car available, P_0

Accessibility variable	Value of accessibility variable	Under £1041	£1041 - £2080	£2081 - £3120	£3121 - £4160	£4161 - £5200	£5201 - £6240	£6241 - £7800	Over £7800	All incomes
All	All	93.8	71.1	42.2	29.5	21.7	13.9	9.4	5.9	54.1
	1	92.3	64.2	37.3	24.0	14.2	5.9	5.8	0.0	46.9
	2	94.9	71.5	33.2	27.3	20.2	5.3	9.3	4.9	49.2
	3	95.1	71.3	46.5	28.1	24.2	16.7	12.8	6.2	55.3
	4	93.0	77.7	52.9	40.9	33.9	38.5	10.3	25.0	65.4
	1	92.7	71.9	48.7	31.1	26.5	16.9	14.3	13.9	58.0
	2	95.2	72.7	44.8	36.5	21.6	26.6	14.7	3.4	57.9
	3	93.7	72.7	41.7	30.4	21.9	8.5	7.7	6.2	54.1
	4	93.5	66.4	33.8	20.6	17.8	8.0	2.0	1.8	46.4
	1	91.7	76.4	46.3	36.0	11.8	-	-	-	60.2
	2	75.0	86.1	66.7	47.4	23.8	-	-	-	63.9
	3	100.01	78.7	56.8	43.2	42.9	-	-	-	61.7
	4	82.4	77.8	55.8	37.4	24.3	20.0	7.7	-	53.7
	5	-	72.0	42.1	34.4	8.2	20.0	0.0	-	40.5
	6	-	51.6	28.9	21.9	8.3	18.7	-	-	28.7
	7	-	33.3	28.3	10.3	0.0	4.2	0.0	8.3	14.4

Table 5

S2	G	ACC band 1	ACC band 2	ACC band 3	ACC band 4	All ACC bands
Household type			Estimates of b_0 (the constant)			
1	2.30	1.72	2.67	3.80	2.72	
2	2.02	2.84	2.88	3.06	2.80	
3	2.29	2.41	3.04	1.74	2.70	
4	2.23	2.50	2.76	3.64	2.97	
5	2.06	1.55	1.84	1.87	1.99	
All	2.42	2.46	2.78	2.94		
		Estimates of b_1 (the income coefficient)				
1	-1.56	-0.78	-1.88	-3.07	-1.81	
2	-1.52	-2.33	-2.26	-2.56	-2.25	
3	-1.70	-1.86	-2.35	-1.45	-2.12	
4	-1.85	-2.03	-2.33	-3.10	-2.49	
5	-1.60	-1.24	-1.52	-1.62	-1.61	
All	-1.87	-1.91	-2.20	-2.45		
		Estimates of income (in £) at which $P_0 = \frac{1}{2}$				
1	3926	20989	3439	2235	4121	
2	2753	2160	2425	2022	2284	
3	2854	2579	2572	2047	2427	
4	2094	2194	1979	1930	2025	
5	2521	2334	2089	1838	2233	
All	2544	2531	2353	2063		

REFERENCES

[1] Department of the Environment, Standard forecasts of vehicles and traffic (1975) DoE Directorate General Highways Technical Memorandum H3/75.

[2] Tanner, J.C., Forecasts of vehicles and traffic in Great Britain (1974), Transport and Road Research Laboratory, Report 650.

[3] Tanner, J.C., Car ownership trends and forecasts (1977), Transport and Road Research Laboratory Report 799.

[4] Department of Transport Report of the Advisory Committee on Trunk Road Assessment (1978) (HMSO).

[5] Bates, J., Gunn, H., and Roberts, M., A disaggregated model of household car ownership (1978), Department of Transport Research Report 20.

[6] Pearman A.D., and Button, K.J., "Regional variations in car ownership" Applied Economics, Vol 8.

[7] Dagenais, M.G., "Application of a threshold regression model to household purchases of automobiles" (1975), Review of Economics and Statistics.

[8] Dumphy, R.T., "Transit accessibility as a determinant of automobile ownership" (1973), Highway Research Record 472.

[9] Shepherd, L.E., "An econometric approach to the demand for urban passenger travel" (1972), Australian Road Research Board Proceedings, Vol 6.

[10] Fairhurst, M.H., "The influence of public transport on car ownership in London" (1975), Journal of Transport Economics and Policy, Vol 9.

[11] West Yorkshire Transportation Studies, Household interview survey report (1976), WYTCONSULT Document 511.

[12] Fowkes, A.S., Initial investigation of the WYTCONSULT household survey data for illustrating methods of car ownership forecasting (1977), Institute for Transport Studies, Working Paper 96.

[13] Button, K.J., Category analysis and household multiple regression models of trip generation: a possible reconciliation (1976), International Journal of Transport Economics, Vol 3.

[14] Fowkes, A.S., Pearman, A., and Button, D.J., The sensitivity of car ownership of households in West Yorkshire to household accessibility (1978).

[15] Cox, D.R., The analysis of binary data (1970), Methuen.

The measurement of travel demand and mobility

16

Yacov Zahavi   United States

SYNOPSIS

This paper reports on some recent developments in urban transportation models, initiated by the World Bank and currently under further development for the US Department of Transportation and the Ministry of Transportation of West Germany. It is pointed out that simplified versions of complex models usually fail as long as the sophisticated models are not fully satisfactory, and it is suggested that simplified models should be based on new approaches to travel demand analysis, rather than follow in the steps of conventional models. One such example is discussed in the paper, where travel demand is represented by the total daily travel distance per traveller/household, as generated under explicit constraints, instead of travel demand being expressed by trip rates. The paper also suggests a quantified definition of mobility, and shows how the new measures of travel demand and mobility can be applied in the evaluation of changing travel conditions, as in Singapore's 1975 before and after studies. The paper concludes with some comments on the technical aspects of data collection to monitor the effects of new transport policies or investments.

SIMPLIFIED MODELS

In response to the requirements of policy makers, transportation models tended to become increasingly complex during the past decade. Of special concern was the models' lack of accurate responsiveness to such issues as mode choice under new rapid transit systems, road pricing in congested areas, and the effect of fuel shortages on travel patterns and urban structure.

In an effort to develop sensitive models, each travel component was analysed thoroughly and simulated by a detailed model, for example for car ownership, trip generation, trip distribution, modal choice and trip assignment. Each model was further divided into sub-models, such as separate trip generation models for different trip purposes. With so many separate sub-models, the interrelationships between travel components received less emphasis, and the complete travel picture tended to be obscured by the over-sophistication of the components. For instance, the close interdependence of the daily trip rate and the average trip distance in a city was lost sight of. It is therefore perhaps somewhat paradoxical that, concurrent with the development of such complex models to meet users' requirements, there was an increased demand from the same users for simplified models, rapid and easy to apply. Indeed, most of the initiative and encouragement for developing simplified models came from such authorities as the Organisation for Economic Cooperation and Development (OECD), the National Cooperative Highway Research Programme (NCHRP), and the World Bank.[1] [2] [3]

The search for new and better transportation models is still going on, and attention has gradually been shifting from the development of new techniques to the exploration of new approaches. One such new approach is described in this paper, with special emphasis on the relative simplicity of its application under changing conditions, such as before-and-after studies.

For reasons of space and because the new approach is still under development, only its general characteristics are described, it being assumed that the reader is familiar with the components and operational phases of conventional transportation models. The new approach was initiated by the Urban Projects Department of the World Bank, and is now under development for the US Department of Transportation and the Ministry of Transportation of West Germany.[3] [4] [5]

TRAVEL DEMAND AND MOBILITY

Conventional transportation models, whether aggregate or disaggregate, recursive or simultaneous, are based on four principal phases, namely trip generation, trip distribution, trip modal choice, and trip assignment. The emphasis in these four phases is on trips, as trips are assumed to represent travel demand. But do they? On one hand, it is obvious that trips reflect the demand for travel, and a trip will be generated only when the utility or benefit of making the trip to a certain destination for a certain purpose surpasses its disutility, such as its cost in time and money terms. On the other hand, it appears that trips alone do not represent travel demand. Consider a traveller who is observed to make six trips a day in his small home town, but is found to make only three trips a day after moving to a large city. Did his travel demand decrease? Is his utility of travel and mobility less in a large city than in a small town?

Let us now add another travel component to the above example, namely trip distance, and we find that the trip distance is 5 km in the former case and 10 km in the latter case. It now becomes evident that the traveller covers 30 km a day in both cities, that he trades off trips versus trip distances. Put another way, city size, which affects trip distances, can also affect trip rates, while the total travel demand, if measured by the daily travel distance, may remain unchanged by city size.

Table 1 presents one such example from England, for the relatively small town of Kingston-upon-Hull and the large metropolitan area of London. Although the size of the study area of Kingston-upon-Hull is only 4.4 per cent that of London, the daily travel distance per average car is similar in both cities, while the trip rate and trip distance are inversely related within the total travel distance. The same patterns were also observed in a wide selection of cities in the US, Europe and developing countries.[6]

An interesting result in the above example is that the same inverse relationship also applies to the trip rate, i.e. trip time within the total daily travel time per average car which, once again, is very similar in both cities, namely 0.72 and 0.75 hours.

One possible interpretation of the above example is that the daily travel demand per average car driver measured by the total daily travel distance, for which the traveller has to pay in both time and money, is practically the same in both cities. Put another way, if we wish to have a measure of travel demand which can be generated within certain daily quantities of allocated travel time and money and which is transferable between cities of different size, it appears that the total daily travel distance per representative traveller is a better measure than the daily trip rate.

It is suggested, therefore, that travel demand should be measured by the daily total travel distance, by all modes, per representative traveller. It is shown later in this paper that this measure of travel demand allows us to unify travel demand, system supply, car ownership and urban structure within one operational framework.

The measure of mobility is currently ambiguous, as it is often used to express a qualitative impression of travel conditions. Mobility is often also used as a synonym for accessibility, as it is evident that the former should be directly related to the latter.

The new definition of travel demand by travel distance allows us to define mobility in quantified terms. It is suggested that it be defined as the product of the daily travel distance per representative traveller/household and the traveller's daily average speed.

Three independent sources corroborate this definition. The first is the Alpha relationship[7] which was derived from empirical analyses òf vehicle flows versus speed on road networks, and can be expressed as:

$$\alpha = qv = Cv^2 \; ; \tag{1}$$

where q = flow of vehicles per unit time along a unit length of road; C = concentration of vehicles per unit length of road; v = space-mean speed; and α = a measure specific to a road section of specific category. For arterials per day, $\alpha \sim 400,000$; for expressways of four lanes, $\alpha \sim 2,500,000$.

The above relationship was also derived from theoretical considerations and can be interpreted in terms of the "kinetic energy" or "kinetic capacity" that a road section can carry.[8] Multiplying the alpha value for a road section by the total length of the road network results in the product of the total vehicle travel distance per unit time (hour or day) and speed.

The second source is a theoretical development of a measure of mobility for a bus system.[9] The need for the mobility measure to satisfy certain constraining conditions leads to the definition of mobility of a bus system as the product of passenger-kilometres and speed.

The third source is the "equilibrium assignment", developed on Wardrop's first principle which can be stated as follows: find the assignment of vehicles to road links so that no traveller can reduce his travel time from origin to destination by switching to another road.[10] One possible way of solving this problem is by minimising a convex objective function subject to several explicit constraints.[11] The mathematical solution of this problem is the minimisation of the area under the link congestion function, which is expressed by the average travel time per unit distance for a given range of flows. It can then be shown that this minimisation process also equals the maximisation of the product of flow and speed. Extending this definition to the total road network expresses the maximisation of the product of travel distance and speed.

Thus, three different approaches lead to the same measure of travel productivity, which is defined here as mobility. This suggests that the result is not coincidental. Furthermore, travel distance becomes the common denominator of travel demand and mobility. The implications of these results for travel demand modelling are far-reaching, as pointed out below.

TRAVEL DEMAND UNDER CONSTRAINTS

Most travel choices are made under constraints, such as time and money. Such constraints are implicit in the mode choice models for single trips, and they can be extended to be explicit constraints on the total daily travel. For example, the daily time and money spent on travel by representative travellers/households are applied as explicit constraints in various models, such as those based on entropy maximisation or utility maximisation.[12] [13] All these models are still based on trips, however, and have to be calibrated to the observed trips in each city. Thus, such models are not fully transferable between cities.

Another approach is to maximise the utility of travel opportunities under the explicit constraints, where the daily travel distance is regarded as a measure of the spatial distribution of opportunities that can be reached within the constraints. This approach is applied in the Unified Mechanism of Travel (UMOT), in which the conventional phases are reversed.[3] [4] [5] The conventional models start with trip generation and conclude with passenger and vehicle kilometres of travel, while the UMOT process starts with the total daily travel distance that can be generated within the travel constraints, and concludes with trip rates.

Perhaps the best way of explaining the process is by interpreting the relationship between the daily travel distance per representative travelled and the door-to-door speed. Figures 1, 2 and 3 show such relationships in two US cities in two periods; the Nuremberg region in West Germany; and Singapore respectively. It is to be noted that in these cases "traveller" refers to a person who made at least one motorised trip during the survey day.

In the first two cases travel is by motorised modes only, while in Singapore the data also include walking trips by motorised travellers. The data for Nuremberg also include regional travel. These relationships are summarised in Table 2. From these we can also derive the relationships between the daily travel time per representative traveller and the door-to-door speed. These are shown in Figure 4.

The results can be interpreted in the following way:

□ When speeds increase, travellers are observed to travel further <u>and</u> save real time. Part of the time saved is traded off for more travel distance (induced travel), and part is saved in real time terms.

□ The saving in the daily travel time is especially significant when low speeds are increased, for instance when travellers transfer from bus to car. This trend may account for the reluctance of car travellers to transfer back to a bus for they then have to spend more time to travel less distance.

□ At relatively high speeds, the daily travel time is relatively stable, most of the time saved being traded off for more travel distance.

□ When extrapolating the relationships to the speed of walking, about 4.7 kph,[14] all converge to about 1.5 hours per representative traveller per day. Thus, the range of daily travel time is relatively narrow, from a maximum of about 1.5 hours to a minimum of about 1.0 hour.

It should be noted that the above relationships represent favourable conditions, where the travellers have the freedom to adjust their travel within their constraints. Under unfavourable travel conditions, however, such as those observed in some cities of developing countries where the poor reside at the fringe of the urban area and have to travel far to job locations, the minimum travel distance can become the binding constraint, and under such conditions the travellers may have to spend as much time and money as required in order to reach jobs. Hence, the distributions by income and location of the daily travel time per traveller strongly influence urban travel patterns.

Another point to note is that the coefficient of variation (standard deviation over mean) of travel times and distances is very similar for all groups of representative travellers when stratified by such factors as income and location.

The second principal constraint on travel is money. The data available from the US suggest that a representative household is willing to spend on travel a maximum of about 11 per cent of income, or slightly more if expressed in terms of disposable income.[15]

About the same proportion was also noted in Canada and England. Furthermore, this proportion is stable for households owning cars at all income levels. Households not owning cars, on the other hand, spend less than 11 per cent of their income on travel. One possible explanation for this trend is that the travellers of such households spend their daily travel time budget long before they reach even six per cent of their income.

Travel demand by distance can now be derived within the two constraints. The following is a simple example, where there are only two modes, bus and car, and two constraints. (When the number of modes equals the number of constraints, there is no need to maximise an objective function, as the problem can be stated uniquely in terms of "n" equations with "n" modes.)

If we define x_1, x_2 as daily travel distance by car and bus respectively; t_1, t_2 as time per kilometre of car and bus respectively; c_1, c_2 as cost per kilometre of car and bus respectively; T as daily travel time budget per household; and M as daily travel money budget per household; then:

$$\left. \begin{array}{l} x_1 t_1 + x_2 t_2 = T \\[1em] x_1 c_1 + x_2 c_2 = M \end{array} \right] \tag{2}$$

Solving the two equations with the two unknowns gives the following outputs: total daily travel distance per household, $x_1 + x_2$; modal split by distance, x_1 versus x_2; time and money allocated to each mode, $x_1 t_1$; $x_2 t_2$; $x_1 c_1$; $x_2 c_2$; average speed, $(x_1 + x_2)/T$; and the measure of mobility, as the product of $(x_1 + x_2)$ and average speed. By this method, travel demand by distance, modal choice and mobility are all direct outputs of one process.

Table 3 summarises the time and money budgets and unit costs of travel for representative households, by income, in Washington, DC, 1968, as well as the resulting demand for car and bus travel distances. The estimated values are shown in Figure 5 as continuous lines versus the observed values, where a close similarity is observed. By contrast, conventional trip generation models have to be calibrated to the observed trips. Hence, their ability to reproduce the same observed trips does not guarantee their validity, especially for forecasts. The ability of a model based on constraints to reproduce observed travel characteristics independently provides a better assurance for its validity. Figure 6 shows the results of an exercise, based on Table 4, for three modes - walking, bus and private modes - and two constraints. The results are of special importance, and suggest the following:

□ There is an income threshold below which a representative traveller cannot afford even a bus fare on a regular basis, as is the case in cities of some developing countries. After this first threshold is crossed, the traveller starts to use bus travel. After a second income threshold is crossed, the traveller starts to use private modes, such as motor-cycles and cars.

□ While walking and travel by private modes are monotonic, decreasing and increasing consistently with income, travel by bus reaches a maximum value at a certain income level, after which it declines. This trend, although known from actual experience in all cities, is now explained by the UMOT process. The analysis also suggests that the key to a better public transport system that will attract passengers from the private modes is speed rather than low fares.

The last subject in this section is car ownership. Conventional transportation models assume that car ownership is a principal factor in trip generation. Hence, outputs from a car ownership model are introduced as inputs into the trip generation model. In the UMOT process, travel demand, car ownership and transportation system supply interact by a feedback process according to the following steps:[4]

□ The travel demand process results in the demand for car travel distance, as one of the outputs.

□ The demand for car travel distance generates car ownership, in order to satisfy the demand.

□ The interactions between the estimated number of cars and a given road network result in unit costs of travel, in time and money terms.

□ The new unit costs are then fed back into the travel demand phase. The sequence is repeated until, by a process of iteration, the total transportation system approaches equilibrium between demand and supply.

□ It should be noted at this stage that in the conventional models travel demand is expressed by trips, while the output of system supply is expressed by the passenger and vehicle kilometres of travel at certain speeds. Hence, in such models demand and supply have no common denominator. In the UMOT process, on the other hand, the travel distance is a common denominator of both the demand for, and the supply of, travel, and the iteration process described above searches the equilibrium point between the two. The travel system is then found to converge rapidly.

Figure 7 shows an example of the car ownership estimation process, based on the demand for car travel distance, as detailed in Table 3; the distributions of travel distance about the mean distances (coefficient of variation is 0.5); minimum daily travel distance that would justify the purchase of cars (14 and 55 passenger-kilometres for one and two or more cars per household respectively); and the probability of a household owning a car at at given income level versus the observed levels is summarised in Table 5, where a close similarity between the two is observed.

In conclusion of this section, a new approach to travel demand is presented that allows us to analyse travel conditions and evaluate alternative policy options in a rapid and consistent way. The implications of this approach for travel conditions in an actual case are discussed in the following sections.

BEFORE-AND-AFTER STUDIES

The following is a short presentation of part of the travel data collected before and after the introduction of the Area Licence Scheme (ALS) in Singapore's central business district in June 1975, which imposed a fee on each car and taxi with fewer than four persons that entered the restricted zone during the morning peak period. The presentation is general and macro in scope, referring to part of the travel data collected from a sample of about 2000 vehicle-owning and non-vehicle-owning households interviewed both before and after the introduction of the ALS. Full details of the surveys are given in a World Bank staff working paper.[16] The bank's paper concentrated mainly on selected trips that were either destined for the restricted zone or directly affected by the ALS, while the following presentation is based on total travel in the Singapore area.

As the original data do not include sample expansion factors, more emphasis is put on the following relationships by income than on the total averages. Furthermore, when comparing the before-and-after data, the following points should be noted:

□ The ALS was introduced during an economic slowdown; therefore, the differences cannot be specifically attributed to the ALS. For instance, it is noted in the World Bank paper that shopping trips by members of vehicle-owning households to destinations in the restricted zone fell by 25 per cent, compared with a decline of 14 per cent to destinations outside the zone. But even the difference in the percentages cannot be attributed to the ALS alone, as a general decentralisation process was in progress in Singapore during the ALS period. Therefore, the World Bank paper is careful in noting that the specific effects of the ALS could not be isolated explicitly.

□ While the flows of cars entering the restricted zone fell by about 70 per cent during the morning peak period when the restrictions were imposed, no significant change was noted in the traffic flows of cars leaving the restricted zone (including through traffic) in the afternoon period. Hence, while the ALS restrictions had a direct and impressive effect on traffic flows during the morning hours, its total daily effects could not be identified and quantified. It seems that car travellers only shifted their time of travel without changing their mode. For example, of about 10,000 parking spaces prepared around the restricted area in the hope that car drivers would transfer to a bus shuttle service to the restricted zone, only about 300 spaces were in fact used. Put another way, car drivers appear to have adjusted their travel patterns in ways other than expected.

The results of the sampled before-and-after home interview studies could not be fully verified because of the lack of recorded data on traffic speeds and bus passenger loadings. Despite these weaknesses, the results reveal several intrinsic characteristics of travel behaviour. Owing to pressure of space, only several selected relationships, stratified by income alone, are shown. Therefore, part of the variability in the trends may be the result of factors such as household size and location, as well as sampling errors. The trends are per average traveller, and the travel time is the reported door-to-door time. The comparisons are divided into households owning and not owning a private vehicle, and before and after the introduction of the ALS.

Figure 3 related the daily travel distance per traveller of vehicle households to that of non-vehicle households versus the door-to-door speed. The before-and-after points intermingle, suggesting that the basic travel behaviour remained unchanged.

Figure 8 shows the daily trip rate per traveller, which is very low, barely over the minimum of two trips. Of special concern is the result that the trip rate per traveller of vehicle-owning households is similar to the trip rate per traveller of non-vehicle households. It is not clear at this stage whether these low trip rates are due to the under-reporting of trips, or are a characteristic of travel in Singapore. The high trip rate per average car, about five trips a day, derived from a comprehensive study suggests that the former explanation is more probable. The daily trip rates of all travellers appear to have decreased during the ALS period, especially for travellers in high-income households. If such trends are found in a survey, it is advisable to recheck the original data carefully in order to identify their causes.

Figure 9 shows the daily travel distance, by car and public transport, per traveller of vehicle-owning households. The data also include the daily travel distance by motor-cycles, bicycles and walking, but these distances are relatively short. The first striking result is that the general trends of the relationships follow those of the exercise shown in Figure 6, namely that while car travel increases monotonically with income, transit travel reaches a peak, after which it declines. While no data on travel costs in money terms or on the number of travellers per household are detailed in the available tabulations, the travel pattern suggests that travel distance is affected by both speed and income, as discussed above.

The second striking result is that the daily travel distance by car declined at all income levels, while only a slight shift to bus travel is noted. While this trend cannot be attributed solely to the ALS, the shift is significant and consistent. Whereas the loss for car travel is great, the gain for transit travel is small, corroborating the indication that modal shifts should take into account the gain or loss of travel and mobility within the daily constraints. This basic phenomonon can be seen in Table 6, where the above data are compared for travellers of vehicle-owning households.

It appears that the loss of travel distance by transfer from car to public transport is 5 per cent; and the total loss by all modes is 3 per cent. It is also shown that the transfer from car to transit by travellers of vehicle-owning households cannot be attributed to the ALS, as the daily travel distance by public transport per traveller of non-vehicle households decreased by about 5 per cent as well. Hence, the most plausible explanation for these results, affecting vehicle and non-vehicle households alike, appears to be the general economic slow-down during the ALS period.

Figure 10 shows the mobility of travellers of vehicle and non-vehicle households, and Table 7 summarises the relevant data. (The data in this case include travel by all modes.) While no major change is noted for non-vehicle households, the mobility of vehicle-owning households dropped significantly.

According to the proponents of road pricing, one of the main advantages is that it allows travellers at all income levels to increase their mobility. Because the mobility of vehicle-owning households actually decreased consistently at all income levels, it may be inferred that this gain was not realised in practice. This may have been due to the level of the ALS fee, which has resulted in the streets within the zone being under-used during the hours of restraint. Non-vehicle households, on the other hand, were not affected much by the ALS fee, thus corroborating the indication that non-vehicle households are affected more by speed than by travel costs, especially at medium and high levels of income.

It can be concluded that a before-and-after comparison based on travel demand by trips, as shown in Figure 8, is not illuminating, as trips express only part of travel behaviour. While under-reporting of trips can affect significantly the trip rate, it affects travel distance and time to a lesser degree, as unreported trips tend to be short. When the comparisons are based on travel distance, speed and mobility, however, they reveal many new aspects and trends, and allow a better evaluation of the travel characteristics in an urban area.

While details of the data required for applying the UMOT process are now under preparation for the World Bank,[17] they can be summarised as follows, especially for before-and-after studies:

□ A relatively small sample of vehicle and non-vehicle households, at different income levels, should be interviewed at both periods. Preferably, travel diaries should be recorded for more than one day in order to decrease the traveller household daily variations of travel constraints.

□ The relationships between daily travel distance and time per traveller versus door-to-door speed, as well as the daily travel distance, time and cost, by mode, versus income, should be derived.

□ Flow-speed measurements, as well as public transport passenger counts, should be carried out at key points of the road network. These data will furnish the values for the Alpha relationship, as well as independent checks of the passenger and vehicle kilometres of travel and mobility measurements derived from the home interview surveys.

Trip purpose should be ranked by trip time and trip distance, in order to derive the ranking of the perceived values of trip purposes. An expected, or planned, change in speed could then affect not only the daily trip rate, but also the proportions of trip purposes within the trip rate.

In summary, the total daily travel distance, time and cost per traveller and per household are essential data for two principal reasons: first, to serve as controlling totals for the analyses and forecasts fo travel behaviour and, secondly, to serve as the parameters for the total travel system within which all the single travel components, such as trip rate, trip distance, trip time, proportions of trip purposes and car ownership interact with one another by trade-offs.

ACKNOWLEDGMENT
I am grateful to E.V.K. Jaycox and the Economic Development Institute of the World Bank, R.W. Crosby of the US Department of Transportation, and Dr H.P. Weber of the Ministry of Transportation of West Germany, for permission to refer in this paper to results of studies conducted for their organisations; and to G.J. Roth of the World Bank for his valuable comments. The views expressed in this paper are those of the author, however, and not necessarily those of the above-mentioned organisations.

REFERENCES

[1] Urban traffic models: possibilities for simplification (1974), Organisation for Economic Cooperation and Development.
[2] Quick-response urban travel estimation: manual techniques and transferable parameters (1977), prepared by Comsis Corporation for the National Cooperation Highway Research Programme, Transportation Research Board.
[3] Zahavi, Y., The UMOT Model (1976), Urban Projects Department, the World Bank, draft.
[4] Zahavi, Y., The UMOT Model: further development, In preparation for the Research and Special Programmes Administration, US Department of Transportation.
[5] Zahavi, Y., and Kocks Consult GMBH, The UMOT Project, In preparation for the Ministry of Transportation of the Federal Republic of Germany.
[6] Zahavi, Y., Can transport policy decisions change travel and urban structure? (1978), PTRC Summer Annual Meeting, University of Warwick.

[7] Zahavi, Y., "Traffic performance evaluation of road networks by the
     alpha-relationship" (1972), Traffic Engineering & Control.
[8] Drew, D.R., and Keese, C.J.,
     Freeway level-of-service as influenced by volume and capacity characteristics (1965),
     Highway Research Record 99, Highway Research Board.
[9] McLynn, J.M., Heller, J.E.I., and Watkins, R.D., Mobility measures for yrban transit
     systems (1972), Urban Mass Transportation Administration, US Department of
     Transportation.
[10] Wardrop, J.G., Some theoretical aspects of road traffic research (1952), Proceedings,
     Institute of Civil Engineers
[11] Eash, R.W., Janson, B.N., and Boyce, D.E., Equilibrium trip assignment:
     advantages and implications for practice (1979), 58th Annual Meeting of the
     Transportation Research Board, Washington, D.C.
[12] Wilson, A.G., Entropy in urban and regional modelling (1970)
[13] Beckmann, M.J. and Golob, T.F., "A critique of entropy and gravity in travel
     forecasting" (1972), Newell, G.F., ed. Traffic Flow and Transportation".
[14] Transportation and traffic engineering handbook (1976), Baerwald, J.E., ed.Institute
     of Transportation Engineers, page 59.
[15] Zahavi, Y., Travel over time (1978), Federal Highway Administration, US Department of
     Transportation, Washington, D.C.
[16] Watson, P.L., and Holland, E.P., Relieving traffic congestion: hte Singapore area
     licence scheme (1978), World Bank Staff Working Paper 281, Washington, D.C.
[17] Zahavi, Y., Urban travel patterns, Economic Development Institute, World Bank,
     Washington, D.C.

Table 1. Travel characteristics of
Kingston-upon-Hull and London, UK

Characteristic	Kingston-upon-Hull	London
Year	1967	1962
Population	344,890	8,857,000
Area, Sq.Km.	107	2,450
Cars	43,185	1,249,450
Cars/100 Persons	12.5	14.1
Car Trips	238,000	4,119,000
Car Trip Rate	6.25	3.27
Trip Distance, Km.	4.15	7.18
Trip Time, Min.	6.9	13.7
Car Daily Distance, Km.	25.9	23.5
Car Daily Time, Hrs.	0.72	0.75

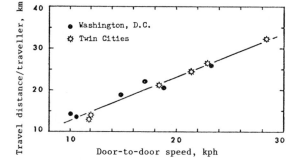

Figure 1. Daily motorised travel distance per
traveller, by mode, v. door-to-door speed,
Washington, D.C. 1955 + 1968 and Twin Cities
1958 + 1970

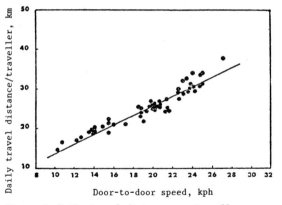

Figure 2. Daily travel distance per traveller,
by household class, v. door-to-door speed, Nuremberg Region 1975

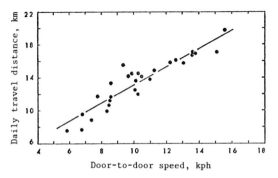

Figure 3. Daily travel distance per traveller v.
door-to-door speed, Singapore before and after study

Table 2. Travel characteristics of travellers v. door-to-door
speed, based on the function: travel distance = a + b(speed), km

Characteristic	Washington, D.C. + Twin Cities	Nuremberg	Singapore
No. of Observations	171	55	28
a	2.18	1.76	2.01
b	1.03	1.20	1.10
R^2	0.87	0.92	0.86
Maximum Time at Walking Speed, hr.	1.50	1.58	1.53
Distance/TR at Walking Speed, km.	7.0	7.4	7.2

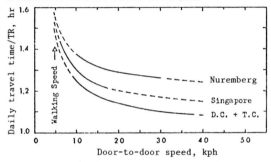

Figure 4. Daily travel time per traveller,
Washington, D.C. + Twin Cities, the Nuremberg region and Singapore

Table 3. Summary of estimated travel demand per household, by income, Washington, D.C. 1968 (times and speeds are door-to-door)

Annual Income, $	4,000	5,000	6,000	7,000	8,000	9,000	10,000	11,000
Cars/HH	–	(0.1)	0.35	0.71	1.02	1.29	1.54	1.76
TM-budget, $	0.51	0.75	1.24	2.01	2.82	3.17	3.53	3.88
TT-budget, hr.	2.02	2.02	2.09	2.20	2.29	2.41	2.53	2.63
CAR: v, kph.	13.5	15.0	16.0	19.0	21.0	24.0	26.0	28.0
c, $/km.	0.104	0.096	0.092	0.081	0.075	0.068	0.064	0.060
D, km.	0.02	2.39	8.40	19.74	34.14	42.38	50.81	60.59
TRANSIT: v, kph.	6.8	7.5	8.0	9.5	10.5	12.0	13.0	14.0
c, $/km.	0.037	0.037	0.037	0.037	0.037	0.037	0.037	0.037
D, km.	13.63	13.96	12.52	11.02	6.97	7.73	7.45	6.56
Total Distance	13.65	16.35	20.92	30.76	41.11	50.11	58.26	67.15

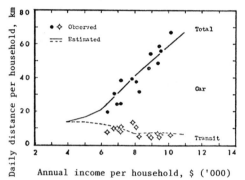

Figure 5. Estimated v. observed daily travel distance per household, by mode, v. income by district, Washington, D.C. 1968

Table 4. Assumed characteristics of the three modes

Characteristic	Walking	Transit	Private Vehicles
Door-to-Door Speed, kph.	4.30	8.60	17.20
Time per Kilometer, min.	13.95	6.98	3.49
Cost per Kilometer, $	0.01	0.03	0.09

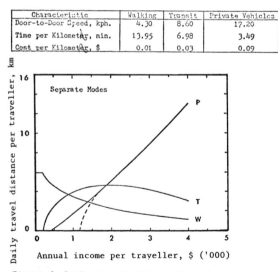

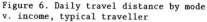

Figure 6. Daily travel distance by mode v. income, typical traveller

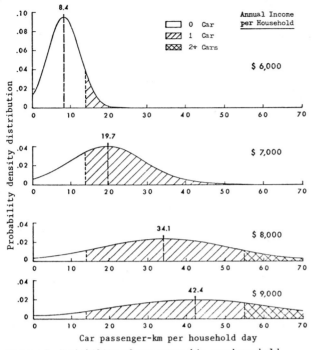

Figure 7. Probability of car ownership per household
v. income, by district, Washington, 1968

Table 5. Car ownership (based on a normal
distribution), by district, Washington, 1968

Annual Income, $	6,000	7,000	8,000	9,000	10,000
Car Pass.Km. (μ)	8.40	19.74	34.14	42.38	50.81
S.D. (σ) = $\frac{1}{2}(\mu)$	4.20	9.87	17.07	21.19	25.41
Prob. 1+ car [1]	0.10	0.72	0.881	0.910	0.926
Prob. 0 Car	0.90	0.28	0.119	0.090	0.074
Prob. 2+ Cars [2]	0.0	0.0	0.111	0.275	0.434
Prob. 1 car only	0.10	0.72	0.770	0.635	0.492
Avg. Car Ownership [3]	0.10	0.72	1.014	1.240	1.447
Observed Car Ownership	0.35	0.71	1.02	1.29	1.54

(1) Assuming 14 pass.km. as the minimum threshold for 1 car;

(2) Assuming 55 pass.km. as the minimum threshold for 2+ cars;

(3) Avg. car ownership = (Prob. 1 car) + 2.2(Prob. 2+ cars).

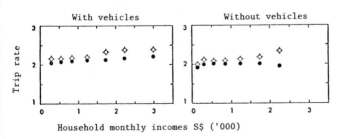

Figure 8. Daily trip rate per traveller v. household
income, Singapore before and after study

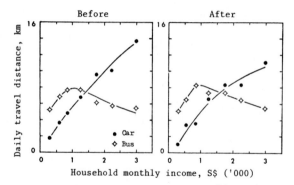

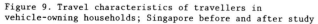

Figure 9. Travel characteristics of travellers in
vehicle-owning households; Singapore before and after study

Table 6. Daily travel distance, by car and
transit, per average traveller of car-owning
households; Singapore before and after study

	Before	After	Change, %
Car Distance, km.	7.75	6.70	- 13.6
Transit Distance, km.	6.71	7.04	+ 4.9
Car + Transit, km.	14.46	13.74	- 5.0
Total, all Modes, km.	16.26	15.27	- 3.2
Speed, kph.	12.84	12.23	- 4.8
Total Mobility	209	187	- 10.5

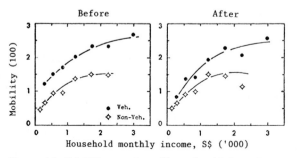

Figure 10. Mobility per traveller of vehicle and
non-vehicle households; Singapore before and after study

Table 7. Mobility per traveller;
Singapore before and after study

Household Monthly Income S$	Owning Vehicle		Non Vehicle	
	B	A	B	A
0 - 200	*	*	45	52
200 - 400	124	83	65	66
400 - 700	152	127	97	91
700 - 1,000	168	140	91	99
1,000 - 1,500	203	195	138	149
1,500 - 2,000	235	230	149	144
2,000 - 2,500	233	207	147	115
2,000 & Over	309	258	*	*

* - Low sample

A method of evaluating urban transportation planning

17

Shogo Kawakami   Japan

SYNOPSIS

In this paper a methodology is developed for the evaluation of urban transportation planning. We should consider the multi-dimensionality of the consequences of transportation systems in the systems evaluation. The consequences include convenience, safety, comfort, cost, and environmental effects. The proposed method is one based on the concept of relative value or utility and disutility. We assume that utilities and disutilities are additive. In terms of this assumption, the utility of a plan can be and is commonly determined by:

$$E_i = \sum_j W_j U_j(S_{ij})$$ (1)

in which E_i = utility of plan as compared with other plans; S_{ij} = outcome state of plan i in jth category of consequences; $U_j(S_{ij})$ = utility of outcome state S_{ij} as compared with other outcome states in the same category of consequences; and W_j = weight of jth category of consequences as compared with other categories. The primary function of the equation is to transform all consequences into value-commensurate units so that they can be unified into a single numerical value to represent the utility of a plan. The plan with the largest utility is regarded as the best.

   The following method to determine $U_j(S_{ij})$ and the weight W_j in the residents' and users' transportation system evaluation is proposed. First, a questionnaire survey is conducted to determine residents' and users' evaluation of transportation systems. Secondly, we determine the value of $U_j(S_{ij})$ which reflects the relative desirability of a category of the outcome states of a plan in the same category that residents and travellers evaluate. Thirdly, we determine the value of W_j which reflects the relative importance of a category of consequences, compared with that of other categories, by using Thurstone's interval scale in the application of the method of paired comparisions used in psychometry in the analysis of the above survey yields. This method can provide more valuable evaluations of transport systems by residents and travellers than methods previously used.

   A method for the optimum planning of transportation systems is proposed.

INTRODUCTION

The process of urban transporation planning most commonly used at present consists of a study of actual transportation states, forecast of land use, determination of objectives of transportation planning, forecast of transportation demands, arrangements of transportation facilities and evaluation of the planning. Although the evaluation process in urban transportation systems planning is very important, the techniques for evaluating these planning systems have more weaknesses than those for forecasting transportation demands. Therefore, an improvement in evaluation techniques is suggested in this paper.

As a rule, evaluation of the transportation system should be based on the views of the people most intimately involved with the transportation system: people who use the system and those who are influenced by it. It is, however, very difficult to determine individual views of the value or to find an commonality of views of the value. A method to find the common evaluation of transportation planning by means of a questionnaire survey of people who are asked to evaluate the urban transportation planning is described in this paper.[1]

The urban transportation system consists of both the main and the local transportation system. Each system has a different function and a particular set of facilities. All residents in the city are linked with the main transportation system, while only some of the residents are linked with the local transportation system. A method for evaluating both of these should be developed. An attempt is made here to devise such methods.

CATEGORIES OF CONSEQUENCES OF TRANSPORTATION SYSTEM

The urban transportation system should be evaluated from the viewpoints of convenience, safety, comfort, cost, environmental effects and space required for the facilities. In a general process of evaluation, an alternative planning procedure is evaluated first with reference to each viewpoint. Next the evaluations from each viewpoint are summed up in terms of the relative weight of each viewpoint. The summed evaluation can show which of the alternative plannings is the best.

In the evaluation of the transportation system, the contents of categories of the consequences of the transportation system should be shown in concrete terms. To understand the evaluations of people objectively, we should indicate the consequences of the transportation system in terms of objective indices: time, distance and cost, and study the people's evaluation of each consequence. Some consequences may consist of concrete indices. For example, convenience consists of the attributes of travel time, walking time, waiting time and transfer times. Therefore, it is difficult to indicate the state of the convenience by only one index. Then, we assume that the evaluation or utility of the consequence of transportation planning can be obtained as a linear weighted combination of utilities of the constituent attributes. Following this assumption, the utility of the convenience can be given as a linear weighted combination of utilities of travel time, walking time, waiting time and transfer times.

EVALUATION OF EACH CONSEQUENCE OF TRANSPORTATION PLANNING

To evaluate a consequence of transportation planning, we should determine the extent of the desirability or the utility which people ascribe to it by experiencing the consequence. We use value transformation functions which are commonly used to transform quantitative outcome states into utilities. The utility functions are shown in Figure 1. We use a 0-to-1 scale for measuring utility. The utility is 0 for the least desirable outcome state and 1 for the most desirable one in the utility functions of all categories of consequences.

The value transformation function or utility function can be derived from the questionnaire survey of people who are involved in the planning. If we ask people whether they are satisfied with the consequence of a transportation system or not, we can obtain the percentage of people who are satisfied or dissatisfied. This percentage will represent the utility or the disutility of the consequence. In order to obtain the utility function, we must collect the data from people in various stages of experience of the consequence of the transportation system.

SYNTHESIS OF EVALUATION OF CONSEQUENCE
In transportation system evaluation we must resolve the multi-dimensionality of the consequences of systems. As stated before, the consequences consist of convenience, safety, comfort, cost, environmental effects and space for facilities.

The proposed method for resolving this problem is an approach based on the concept of relative value or utility and disutility. We assume that utilities and disutilities are additive. On this assumption, the utility of a plan or transportation system can be and is commonly determined by:

$$E_i = \sum_j W_j U_j(S_{ij}) \qquad (1)$$

where E_i = utility of plan i as compared with other plans; S_{ij} = outcome of plan in in jth category of consequence; $U_j(S_{ij})$ = utility of outcome state S_{ij} as compared with other outcome states in the same category of consequence; and W_j = weight of jth category of consequences as compared with other categories.

The primary function of Equation 1 is to transform all consequences into value-commensurate units so that they can be unified into a single numerical value to represent the utility of a plan. The plan with the largest utility is regarded as the best. We can replace the utility with the disutility in the equation. In this case, the plan or transportation system with the smallest disutility is considered the best.

We assume that the relation between the utility of a consequence and the utilities of its attributes can be represented by the following equation:

$$U_j(S_{ij}) = \sum_k w_{jk} u_{jk}(s_{ijk}) \qquad (1')$$

where s_{ijk} = outcome state of plan i in kth attribute of jth category of consequence; $u_{jk}(s_{ijk})$ = utility of outcome state s_{ijk} as compared with other outcome states in the same attribute of the consequence; and w_{jk} = weight of kth attribute of jth category of consequences as compared with other attributes. This equation is essential to the first equation.

METHOD FOR DETERMINING WEIGHT W_j
A method is now proposed to determine the weight W_j in the evaluation of the residents and users of the transportation system. Two different approaches to the method are possible. In the first, Methodology I, we present transportation system plans to the people and study their evaluation of the plans by way of questionnaires. In the second, Methodology II, we study by way of a questionnaire survey the people's evaluations of the actual transportation system which they experience every day. The former method

has the advantage that we can get several responses or evaluations of the various states of a consequence from the same person. The latter method does not have this advantage, but it allows us to acquire a more accurate evaluation by the people of the consequence than is possible with the former, because the people can compare one consequence of the transportation system with another and they can every day experience various transportation systems.

Methodology I. We carry out a questionnaire survey of the residents' and users' evaluation of several proposed transportation systems. The questionnaire consists of three sections. The first measures attitudes to the consequence of transportation systems, the second collects information on the comparison of transportation systems, and the third collects socio-economic information on the respondents.

In the first part of the questionnaire we ask the people whether they are satisfied with each consequence of the system or not. We assume that the percentage of satisfied people represents the relative desirability of the consequence. From the first part of the questionnaire we can determine the value of $U_j(S_{ij})$, which reflects the relative desirability of a category of the outcome states of a transportation system in the same category evaluated by residents and travellers.

In the second part of the questionnaire we ask the people to choose the better of each pair of proposed transportation systems, or we ask them to rank the transportation systems in order of desirability. From the response to this question we can obtain the percentage of people who prefer one transportation system to another. We can then determine the value of weight W_j which reflects the relative importance of a transportation system, compared with other systems, by using Thurstone's interval scale,[2] in the application of the method of paired comparisons used in psychometry in the analysis of the above questionnaire survey yields.

We assume a normal distribution with mean value μ and variance σ^2 for the people's evaluation of the transportation system, E. If the percentage of people who decide that transportation system A is better than system B is p, $0<p<1$, we can estimate that the probability of $E_a>E_b$ is p, or $P(E_a>E_b) = p$.

Therefore, using the symbols shown in Figure 2, we arrive at:

$$\mu_a - \mu_b = C_{ab} \tag{2}$$

where μ_a, μ_b = mean psychological values of people's evaluations of transportation systems A and B.

Since the distribution of E_a-E_b is represented by $N(\mu_a-\mu_b, \sigma_a^2+\sigma_b^2)$, we can determine C_{ab} from the following equation:

$$\int_0^\infty 1/\sqrt{2\pi(\sigma_a^2+\sigma_b^2)}\, \exp\{-(x-C_{ab})^2/2(\sigma_a^2+\sigma_b^2)\}dx = \int_{-q}^\infty 1/\sqrt{2\pi}\exp(-t^2/)2dt = p \tag{3}$$

where

$$q = C_{ab}/\sqrt{\sigma_a^2+\sigma_b^2}$$

In order to determine C_{ab} with Equation 3, we must obtain the values of σ_a and σ_b. If the value of σ_a is generally nearly equal to that of σ_b, we can obtain the value of C_{ab}

with the standard unit $\sqrt{2}\,\sigma_a$. Since we need relative instead of absolute values of C_{ab}, we can assume $\sqrt{2}\,\sigma_a = 1$. If we assume that the relative weight W_j gives the mean value of E, equation 2 gives us

$$\sum_j W_j U_j(S_{aj}) - \sum_j W_j U_j(S_{bj}) = C_{ab} \tag{4}$$

where the value of C_{ab} represents the difference between the mean psychological values of transportation systems evaluation.

By applying Equation 3 to the results of the questionnaire survey on various transportation systems, we can derive much data for Equation 4. Then, by applying the least-squares method to Equation 4 we can derive the relative weight W_j of jth category of a consequence of the transportation system. The reliability of the weight W_j can also be estimated by the statistical method.

Methodology II.[3] We divide the study area into zones in which the consequences of the transportation system are nearly equal. We carry out a questionnaire survey on the residents' and users' evaluation of the actual transportation systems. The questionnaire consists of three sections. The first and third are the same as those used in Methodology I. We determine the value of $U_j(S_{ij})$ from an analysis of the response in the first part of the questionnaire.

The second collects information on a comparison among categories of consequences of the transportation system which the people experience daily. In the second part of the questionnaire, we ask the people to choose the better of each pair of consequence of the actual system, and to state which consequences in each pair they want to be improved. Or we ask them to rank the consequences in order of desirability or undesirability. From the analysis of the response to this questionnaire, we can derive the percentage of people who, considering the actual consequences of the transportation system, prefer either component of each pair of consequences.

We assume a normal distrubution with variance σ^2 for the people's evaluation of the outcome state of transportation system i in jth category of consequences, E_{ij}. If the percentage of people who estimate that jth category of consequences is better than lth category is p, $0 \leq p \geq 1$, we can estimate that the probability of $E_{ij} > E_{il}$ is p, or $P\{E_{ij} > E_{il}\}$ = p. Therefore, using the same procedure as in Methodology I, we obtain:

$$W_j U_j(S_{kj}) - W_l U_l(S_{kl}) = C_{jl}^{-} \tag{5}$$

where the value of C_{jl}^{-} represents the difference between the mean psychological values of jth and lth categories of consequences and k represents the zone. Applying the least-squares method to Equation 5 by using the data from the questionnaire survey on the various zones, we derive the relative weight W_j of jth category of a consequence of the transportation system. Methodology II may be considered a special variation of Methodology I, i.e. Equation 5 is a special case of Equation 4.

Since the psychological values of evaluation of consequences in Equations 4 and 5 are represented as the product of the relative weight W_j by the utilitity of consequence of transportation systems $U_j(S_{ij})$, it is possible to obtain the value of W_j independent of the state of the consequence experienced. That is, by using this method, we may obtain the universal value of W_j.

APPLICATION OF THE METHOD

In the application of Methodology II, we can divide the consequences of the transportation system into two or three categories in order to make it easier for people to compare each pair of consequences. For example, one may comprise positive items, or convenience, safety, comfort, etc, and the other negative items, or envirionment effects, cost, space required, etc. Respondents can easily compare one consequence with the other in each group. In each group we can obtain W_j by using this method. The summed evaluations of consequences in each group are represented as follows.

$$F_k = \sum_j W_j^f U_j^f(S_{kj}), \qquad B_k = \sum_l W_l^b U_l^b(S_{kl}) \qquad (6)$$

where F_k = the summed evaluation in zone k of the group of consequences f; B_k = the summed evaluation of zone k of the group of consequences b; S_{kj}, S_{kl} = outcome states of jth and lth consequences of the transportation systems in zone k; and W_j^f, W_l^b = relative weights of consequences j and l in the evaluation in groups f and b respectively.

The combined evaluation E_k in zone k is represented as follows, on the assumption that the utilities and the disutilities are additive.

$$E_k = W^b B_k + W^f F_k \qquad (7)$$

where W^b and W^f represent the relative weights of B_k and F_k in the combined evaluation.

We can represent the evaluation of consequences of transportation systems by using utility or disutility: satisfaction or dissatisfaction. The application of this method shows that the representation of evaluation by the degree of dissatisfaction makes the questionnaire survey easy and also enhances the accuracy of the data derived from the survey.

Since we can get many values for F_k, B_k and E_k from the questionnaire survey, the values of W_j^f, W_l^b, W^b and W^f can be easily determined.

An application of Methodology II to the evaluation of environmental effects and convenience of an arterial street in Nagoya was attempted and the results of the survey are shown in Figures 3, 4 and 5.

Applicaton of Equation 7 gave the relation of $W^b : W^f = 1.00 : 1.92$.[4] Analysis showed the relative weights of noise, vibration, air pollution, danger, dust, severance, infringement of privacy and encroachment of view of the street were 1.25, 1.05, 1.38, 1.00, 0.84, 0.69, 0.51 and 0.57 respectively.[2] The coefficients of correlation in the regression analysis showed that all these coefficients were significant.

PLANNING OF TRANSPORTATION SYSTEM

In this section a method is described for determining the optimum planning of a transportation system.

□ Study area. The study area should be that in which the people affected by the transportation system mainly live. The study area is divided into zones in which the effects of the transportation system are equal.

□ Estimation of transportation states and their environmental effects. The transportation demands of the system should be estimated and its environmental effects and the relative weight of each consequence of the system W_j should be estimated by using the method described above.

□ Optimisation of transport system. The estimated states of the transport system should be evaluated from the viewpoint of convenience, safety, cost, environmental effects and space required for facilities. Then we should devise minimum standards for convenience, safety, comfort and environmental effects which the transportation system must satisfy.

The systems which satisfy the standards should be considered feasible. The transportation system which reduces the sum of the people's dissatisfaction to a minimum when the cost is below a certain amount should be considered optimum.

This system is represented by the following mathematical equations:

Minimise $$E = \sum_{kj}\sum W_j U_j(S_{kj}) N_{kj}(S_{kj}) \tag{8}$$

subject to $$\sum_{kj}\sum S_{kj} C_{kj} \leqq C \tag{9}$$

$$U_j(S_{kj}) \leqq U_j(S_j), \quad j=1,\ldots,m, k=1,\ldots n. \tag{10}$$

where $N_{kj}(S_{kj})$ = number of people dissatisfied with the state S_{kj} of the jth consequence of the transportation system in zone k; C_{kj} = cost of the state S_{kj} of the consequence of the transportation system; C = maximum value of the resource; S_j = minimum degree of the state S_j of the jth consequence of the transportation system which people can admit; m = number of consequences of the transportation system; and n = number of zones.

CONCLUSION

This paper proposed a method for evaluating transportation systems. The concept of relative value or utility and disutility was used to transfrom the multi-dimensional consequences of a transportation system in the system evaluation into value-commensurate units. Thurstone's interval scale used in psychometry was used to determine the relative weight of the utility of each consequence in transportation system evaluation. A basic method was presented to optimise transportation system planning by using this procedure.

The methods presented in this paper are promising, but more research is needed to understand the relationship between the evaluation of a transportation system and the socio-economic conditions of the respondents, to determine the state of S_j and to investigate the relationship between S_{kj} and C_{kj} in the planning system.

The data used were collected from a small area where there are no complex travel problems. Similar application with larger data sets from larger metropolitan areas should be carried out to revise the methodologies presented in this paper.

ACKNOWLEDGMENT

The author expresses his appreciation to Professor H.G. Retzko for his help in completing this paper, and to Naojiro Aoshima and Kazuo Katahira for their suggestions. He also wishes to acknowledge the financial support given by the Alexander von Humboldt-Stiftung.

REFERENCES

[1] Kawakami, S., and Aoshima, N., A method for evaluation of highway planning (1976), Proc. of Annual Meeting of JSCE Chubu Branch, pages 173-4 (in Japanese).
[2] Guilford, J.P., Psychometric methods (1954), second edition, pages 178-96.
[3] Aoshima, N., Kawakami, S., and Katahira, K., An analysis on significance of components in environmental evaluation systems for highway planning (1977), Proc. of JSCE, No 263, pages 97-106 (in Japanese).

4   Kawakami, S., Aoshima, N., and Katahira, K., <u>Analysis of relation of trading off</u>
    <u>between the convenience and environmental effects in the Evaluation of</u>
    <u>Transportation System</u> (1977), Proc. of Annual Meeting of JSCE Chubu Branch (in
    Japanese).
5   Lin, F.B., and Hoel, L.A., "Structure of utility-based valuation approaches (1977),
    <u>Transportation Engineering Journal</u>, ASCE, Vol 103, No TE2, pages 307-20.

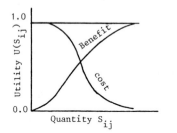

Figure 1. Utility functions

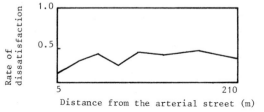

Figure 3. Evaluation of convenience

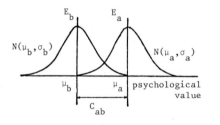

Figure 2. Distributions of evaluations E_i

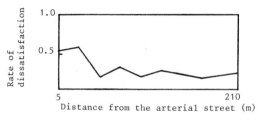

Figure 4. Evaluation of environmental
effects

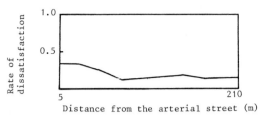

Figure 5. Combined evaluation

Transportation engineering: 18
a rationalised general model

Alessandro Orlandi   Italy

SYNPOSIS

What is a transportation problem and how should it be confronted so that its study will give the most reliable results? This is a question which engineers have asked and are still asking, bringing to the surface an identity crisis which has remained latent for some time, even if it has been manifest clearly only in the past few years. This question can be answered by the formulation of a generalised model of transportation, i.e. a rational approach which is first of all suitable for encompassing and expressing the entire transportation question in a single, disciplined form, and which then puts the engineer or researcher in a position to know the true significance and correct formulation within the general theme.

In this manner it is possible to construct a method for confronting any transportation problem, especially those dealing with the territorial planning of transportation which, extending beyond the immediate contingent aspects in a spirit of interdisciplinary collaboration, makes it possible for new goals in transportation science to be reached.

INTRODUCTION

There are numerous problems with transportation as a basic theme. They include not only those concerned with transportation in the strict sense of the word, but also those concerned with other sectors, be they planning or organisation. Territorial planning of transportation represents a nearly complete synthesis of all sectors involved in transportation. We shall adopt this theme to set out ideas on a topic which is closer to the scope of this volume, to serve as reference for an initial approach to the present study. It is well known that studies of territorial transportaion planning, which have been carried out in a number of cities and regions, have a history of more than twenty-five years. Methods have been well tested not only in theory but also in the application of the processes connected with the various phases of study and planning. Today the logical scheme for territorial planning of transportation is still based on the Detroit and Chicago plans of before 1960, apart from the obvious improvements made in model prediction, while there has also been an effort to perfect the methods of evaluating the choices.

The considerable progress made in simulation processes and the application of individual methods, and the improved knowledge of phenomena acquired through the great amount of statistical information gathered in the meantime and analysed the world over, have allowed us to reach a highly sophisticated level of methodological and simulation perfection which, necessarily, has been enhanced by the evolution and wide use of computers.

The level of perfection which has been reached in all phases of the study of planning - the general procedure, the formulation of models, calibration methods, computational instruments, the wide availability of raw data and means of comparison - can, however, lead to a condition of relative impasse which is the more dangerous the less apparent it is (because it is hidden by abundant model production and better instrumentation), unless we immediately move on to search for a new logical process which, without neglecting previous studies, gives a different but more efficient view of the general question and thus provides new methods of research and work, advancing the scope of our knowledge.

A truly substantial development in research on territorial planning of transportation is possible by identifying a philosophy which can tackle any transportation problem in a complete and unified way. In fact, a new frontier of knowledge cannot be reached if we continue to consider the transportation question in section, i.e. if we continue to think only in terms of territorial planning or the construction of transportation apparatus, vehicle manufacture, and so forth.

We think that transportation must be considered a single phenomenon and confronted as such. This requires that we consider and analyse in a single, disciplined fashion all aspects of transportation. As will be pointed out later, the problems of transportation must first be identified and analysed by a single generalised method and, finally, a basic logical process to study and solve them must be developed.

ANALYSIS OF THE TRANSPORTATION QUESTION

When studying engineering problems of a certain importance, engineers must make their own contributions to be used in different ways or forms essentially concerned with planning and organisation. If we look for a moment at which specialised sectors can make a valid contribution to the transportation question, we see that they are numerous and varied since they may include, for example, structural design, mechanics, hydraulics, electrical engineering, geological considerations, informatics, statistics, and so forth.

It follows that transportation appears to be more of a mixture than a single discipline and, as such, it would lose its interest as the essence of transportation, whereas the territorial planning of transportation as a structuralistic, economic, social or other fact and the planning of transportation systems as a mechanical-design fact, would be of interest.

We believe that, like the other disciplines in engineering, transportation can be a true discipline in itself, even though it has many inter-disciplinary facets, because these appear as specialised sectors which solve their engineering problems or make their own contribution, together with that of other disciplines, to the solution of another problem in which different characteristic sectors appear, among which is transportation. We can therefore pose some questions, the answers to which will tell us whether a transportation discipline is possible.

□ Does transportation constitute a discipline in itself?
□ If it does, of what does it consist?
□ How is it organised and of what is it composed?
□ What logical process must we use to confront it?

We can reply in the affirmative to the first question, for it is possible, as we shall see later, to fit any transportation problem into a single model which will express it in terms of the same basic elements with a single goal.

We can now move on to answer the second question. It is composed of a set of problems which have transportation as a common matrix. Into it come all those problems which have to do with true transportation action, i.e. in which the act of transportation is the immediate scope, such as the organisation of a transportation or distribution service. From this it follows that even the organisation of a transportation company and the planning of the network is a transportation problem. Therefore, the organisation of transportation in a territory, i.e. the territorial planning of transportation, is also a transporation problem.

To put the organisational study of a company or the formation of a territorial transportation plan into effect, it is necessary to know, among other things, the transportation systems available, from both the functional and structural viewpoints. Functional, because it is this characteristic which makes the system valid or not for service; structural, because structure determines the feasibility and practicality of the system.[1]

In conclusion, we can answer the second question with the following syllogistic proposition: since those problems of transportation actions are without doubt transportation problems, the functional aspect is necessarily a transportation problem, if the function of the transportation apparatus (route and/or vehicles) concerns true transportation. But in this case even the structural aspect is a transportation problem, if planning is connected to the function in the manner described above. We have thus demonstrated that even the planning and running of transportation systems fall within this sphere of interest, because they are transportation problems.

With regard to the third question, we can state that any transportation problem includes three essential elements: the scope, the basic components and the field of interest.

□ The scope is the end-point of the work; it may take two directions, planning or organisation.

□ The basic components are the elements which make up every problem and are essentially those basic factors which continually come into play in every transportation problem. They can be identified by a simple logical process: all transportation problems are connected to man, the originator, user and operator of the transportation action; to the object of transportation; to the environment in which the action takes place; to any kind of apparatus employed to carry it out; and to the decision-making process which is necessarily present since every transportation action is the result of a man-made choice.

□ The field of interest is represented by the object of study, which can have as its aim either planning organisation. Considering the transportation question as a whole, including planning and organisation both of transportation systems (or parts of them) and territorial systems (or parts of them), we can classify the problems, with regard to the field of interest, in two categories: one is spatial (the territory being physical, the company being administrative), the other structural (i.e. the transportation system in general and its components, the routes and vehicles).

Finally, with regard to the logical process with which the study of transportation must be confronted, after pointing out that two conditions (safety and regularity) must be met in order to define a spatial transfer in transport, one arrives at the formulation of the basic postulate in terms of which transportation itself must be the result of a real demand previously generated in a given sector. From this postulate is derived the simple and logical method by which what is studied and carried out must be such as to permit the requirements to be met in the existing environmental conditions and with the existing restraints.

A GENERALISED MODEL

In terms of what has been outlined above, the transportation question can be summarised as follows. Knowing the five basic components, one can confront each transportation problem by defining its scope and field of interest.

Before entering into a detailed analysis of the individual elements, and in order to make the general description of the theme as clear as possible, we can illustrate the model by means of the diagram in Figure 1. In each transportation problem there are three logical phases: the scope, i.e. the type of study to be done; the field of interest, which represents the sector in which one must operate; and the basic components, which represent the overall knowledge available to perform the job. The diagram is not organised in terms of time or formal logic, because it puts scope first and field of interest last. One might think the latter should come first, since, in any case, the sector involved is the first thing known. Instead, the sequence can be interpreted as in a logical-functional context: that which one wants to reach is a result identifiable in a given field of interest for which the basic components are used, with full cognisance of the scope. We can now proceed to examine in detail the single elements.

The scope. The scope is the object in sight and is, therefore, the object of the solution to the problem posed. It can entail planning (construction) if it implies the conception, design and calculations for the construction of new entities, identifiable not only for their own physical components but also for their functional characteristics. Or it may be an organisational problem if it concerns researching the functional arrangement of a set of existing physical entities with a view to arrive at a different, necessarily better, use of these entities. These two aspects generally appear together and, one dominating the other, characterise the scope.[2]

The basic components. The basic components are those elements which play a role in the creation of every transportation problem or which may affect it. Every transportation problem necessarily includes the means of transportation which moves a certain object in an environment in order to serve man, who must make decisions on the basis of which the final solution of the problem will be reached. Thus, in the simplest cases as in the most complex, the basic elements which cannot be omitted are: the environment, man, the object of the transportation, the transportation apparatus and the decision-making process. It is obvious that these basic elements are involved in every transportation problem. The basic elements can be defined as follows.

□ The environment. The environment is the physical or relational space within which the question under study is framed, i.e. it constitutes the space in which the event which produces the transportation problem occurs. The physical environment is, in general, a territorial space, with its true physical, relational, socio-economic, historical, cultural, administrative and political characteristics. More precisely, it is a true territory in studies of the territorial planning of transportation, an economic-administrative space in company organisation studies, and a physical place of use of transportation structures in the planning of roads, vehicles and transportation systems in general.

▫ Man. Man enters into the problem in the dual role of user and administrator of transportation. As user, he travels or benefits from the goods carried; as administrator he is the designer, organiser and constructor of the system or services. In these two roles, man appears under different aspects which produce conflicting requirements for the solution of which man again must intervene with the conclusive choice. The interest in man concerns different aspects, such as the number of persons when large masses of people are involved; the extent to which account is taken of the geometric coherence between the structure, passenger and driver; the physical behaviour in response to certain factors arising from the transportation itself (pollution, vibration, acceleration, pressure, etc); and finally, the psychological aspect with regard not only to the driver (perception, behaviour) but also the user who must be considered the consumer of the goods produced, i.e. the transportation service.

▫ Object of the transportation. The transportation system, its design and organisation, constitutes the problem under study and has as its scope the transportation of that which is generally defined as the object of the transportation, which necessarily assumes major importance and affects the study from the very beginning, in both its technical formulation and the decision-making process. The object of the transportation can be people, i.e. entities whose specific aspects were considered in the preceding section, and of things, i.e. inanimate objects and animals.

▫ Transportation apparatus. The term transportation apparatus means any fixed or moving system which has or can have a relation with transportation, whether on land or not. Since the same subject is treated in another section of this paper (Field of interest: structural) it is opportune now to motivate the need to distinguish two stages of study. Whereas in this section (Transportation apparatus) attention is directed towards all the structural entities, including routes, vehicles and plans in general, which are useful for carrying out the transportation, in the other section (Field of interest: structural) the focus is on the functional aspects and the constructive aspects of the transportation system as a real entity. In this manner, the present section (Transportation apparatus) also examines the equipment as long as it is considered useful, even if not actually used in the past in the transportation sector (especially motors, technical short cuts, etc). The second looks at the elements which must constitute an entity or, better, a system suitable for use in transportation. Thus, we can include in the first category an internal combusion engine (which could be used to run pumps and generators as well as vehicles) while the motor can be included in the second category only if it is part of an efficient vehicle.

- Transportation apparatus can be usefully divided into three categories: vehicles (moving equipment), infrastructures (fixed apparatus) and auxiliary installations.

- Vehicles are the mobile elements of the transportation system. They move in suitable locations to carry out the transportation and (real or imaginary) are the true transportation elements.

- The infrastructures constitute the fixed elements of the transportation apparatus and, if they exist, represent the site of the motion.

- Auxiliary installations are all those apparatuses which are neither vehicles or infrastructures. Essentially they can be divided into two categories according to their use: those for power supply (the production, supply and distribution of energy), and those for control (traffic regulation, system protection, etc).

□ The decision-making process. This is a series of operations which are performed and means which are adopted to permit the practical individualisation of the set of structural and relational entities which, satisfying the basic conditions posed by the problem, optimise the target function originally formulated by the person who originated the study. In the decision-making process we can identify the criteria for choice or the criteria on which the target function to be optimised is based; the method adopted to establish what values to assign to the parameters which come into play in the problem and, finally, the instruments which consist of the mathematical and calculations means to elaborate the data with the aim of obtaining the required information.

□ The criteria for choice are based on three basic factors: political, economic and efficiency.

- The political criterion consists of a series of decisions which must be taken by the person who ordered the job. The choice which he makes can also take account, as in effect it does, of other aspects. Then the selection criterion is obtained by the combination of two or three of the above-mentioned basic criteria. In any case, the political component of the choice is always there, if in nothing other than the selection criteria for the criteria for choice.

- The economic criterion is based on a balance expressed in monetary terms between profits and losses, between benefits and costs facing those who manage the transportation. When the values assigned to the problem's independent variables in a certain scheme vary, or the scheme itself varies, or there is a change in management or transportation, the difference or ratio between benefits and costs is maximised and the optimal solution is considered to be achieved. In reality, this decision-making criterion is not always so easily applied, because the procedures applied advantageously in one case are not necessarily valid in another.

- The criterion of efficiency is based on the improvement of part of the system's performance with respect to the energy consumed. In other words, efficiency measures the system's yield, expressed as the ratio between resulting useful power and that consumed. The criterion of efficiency is thus essentially technical, being based on the technical criteria of performance.

□ The method to be adopted for making the evaluations which refer to the economic and efficiency criteria is composed of the entire set of logical and logico-mathematical models which can be grouped under the vast general heading of theory of decisions, and the instruments which it uses, such as operational research and statistical analysis.

□ The instruments to be employed to solve the problem, with the methods selected because they are deemed suitable for making a choice, are those actually used: mathematical instruments (solution algorithms of models in approximation procedures, etc) and instruments for calculation (computers, simulators, etc).

The field of interest. This constitutes the true object of the study, subdivided into two elements, spatial and structural, as shown in the diagram in Figure 1. Grouped in the first are those problems connected with territory and companies which express a concept of space which is physical for territory and managerial for the company. The second groups those problems which implicate a structuralistic concept which may be subdivided into three sectors: route, vehicle and transportation system. These express, in turn, concepts of location alone, vehicle alone, and location and vehicle together.

The division into two sectors, spatial and structural, is not only of formal but also of substantial interest. In fact, when one proceeds to the study of one of these sectors of interest, whether for operation or planning, the approach to the work will be directed according to whether the sector belongs to the first group (complex) or the second (individual). In the first case, the importance of the problem will be characterised by a wide, inclusive vision of the phenomenon, in which a greater effort for synthesis will necessarily be developed. In the second case, it is characterised by a thorough vision which will require a greater effort for analysis.

Significance of the model. The model presented here should be considered a useful instrument for an easier approach to the study of any transportation problem, framing it in the most comprehensive manner possible in the appropriate context. In this manner, one will not only confront and solve the question in terms of the environment and available mobility, with more or less sophisticated techniques, and the instruments available. Instead, one will also take into consideration the present and future technology which is useful for the creation of transportation systems, delving into the socio-economic philosophies which could affect or generate new ways of life and thus new transportation requirements, thereby taking account of man's physical, psychological and social aspects, considering man not only as a component of the community to whom it is possible to attribute an autonomous behaviour, even if determined by that which surrounds him (environment, stimuli, messages).

Analogously to the case of the territorial planning of transportation, the design and production of a particular vehicle or a transportation system, or even a vehicle or transportation system in general, can no longer be confronted without those factors connected to the user and his environment. Contrary to what has generally happened so far, the environment cannot be considered a purely physical fact. It is also a socio-economic fact because it is clear that a given choice in the field of transportation can change the way of life of a population.

Therefore, framing the transportation question in terms of the methodology of the model proposed in this paper facilitates the engineer's study work, because it puts him in a position to formulate the most correct and most complete procedure to confront the practical solution to any transportation problem. It does not limit the study to the more immediate aspects, but researches and analyses also those aspects which, while seemingly unimportant, are actually quite significant. It also permits the researcher to identify the connections between apparently unrelated elements, thus affording the possibility of confronting the study of the single aspects from opposite directions.

This model, which therefore cannot be considered pure speculative philosophy, is important from the didactic viewpoint as it permits one to deal with the entire subject, giving it the nature of a discipline and facilitating a complete coordination of the various contributions of other disciplines.

FORMULATION OF THE LOGICAL PROCESS

In the preceding sections we identified the transportation problems and expressed them in a unified form by means of a generalised model. To complete the formulation of a method for approaching the study of the transportation question, we must first define an approach with which one can confront and solve the particular problem once it has been singled out and identified. Before formulating the logical process in the correct form, there are some general observations to be made.

Transportation represents the act of bringing persons or things from one place to another, thereby carrying out a corresponding transfer. While the transfer constitutes the passive aspect in that it represents the action undergone by the object of the transportation created by opportune apparatus, transportation constitutes the active aspect in that it produces the transfer, but the same cannot be said for transportation. A fall from a precipice can be considered a transfer, but certainly does not constitute a transportation act. It would become transportation if this transfer were carried out for a definite reason. For example, if this transfer (transportation of a wounded person or supplies) is done for a specific reason and with the desire to avoid damage to the object of the transportation, this action would be considered a transportation.

Therefore, to be considered as such, a transportation act must have a certain finality or usefulness for man or must, in any case, constitute an important element in his eyes. The transportation downstream of a tree trunk floating on the water of a river is considered a true transportation act when man takes advantage of this; otherwise it would be an object floating away.

When man assigns a scope to this transfer, it signifies that he has made a series of considerations and decided on its usefulness. For this reason, the transportation act must be programmable, so that one can foresee where, how and when the beginning and end of the transfer occur, as well as the manner of carrying it out.

Two conditions must therefore be met: safety and regularity. The transportation act must take place so that the object of the transportation undergoes no damage resulting in the partial or total loss of its value. Also, one must meet the condition of predetermining the time and place of the journey, i.e. departure and arrival. The first condition refers to true transportation as a phase of connection; the second to the space-time restraints.

Transportation is a cause-and-effect relationship and is generated as a consequence of a desire to carry out a transfer in the desired manner and time periods. Thus, we can express the following basic postulate on the general principle of transportation motives: generation of the transfer demand; carrying out of the transportation and fulfilment of the demand. Obviously, these two points constitute the essence of the phenomenon, because the transportation act has no sense if there is no demand for the transfer. In other words, one can study the most suitable system (as a structure or organisation) after the transfer demand has arisen.

From the postulate derives the methodology to use when confronting the study of any transportation problem: to know the need for movements and the conditions in which it is generated and by means of which it develops; and to study and create the system suitable to make this transfer possible in these conditions. We can now synthesise the entire mechanism of generation, the general study of transportation problems, in terms of the diagram of Figure 2.

There is an actual problem from which, with the proposed method and by means of the generalised model, one can arrive at an expression of a particular model. From this, one passes to the analysis of the phenomenon in respect of the evaluation of the demand, the environment, the restraining conditions and the available technical instruments. After the point of decision making, one finally arrives at the planning and execution of the system.

One encounters considerable difficulties in developing the process discussed. It is erroneously believed that it arises only in problems of the territorial planning of transportation. Actually, analogous difficulties are also encountered in other studies, even if in a less marked and pressing manner. Examples are the optimised planning of transportation systems or simply vehicles (the case of the lunar vehicle comes to mind). The phase of providing for our future world is very delicate (the lunar vehicle consideration was for another world!). One cannot be limited to considering only the numerical values of the evolutionary parameters; we must also take into account their characterisation which may sometimes take on great importance. We can cite two of the most representative examples.

□ The first is the influence that population has on future movements, not only for the quantitative increase but also the greater mobility owing to increased funds and the modification of aspirations.

□ The second concerns the creation of new transportation systems, with implications not only for a population's social and behavioural nature which influences the system's overall conception, but also for the implementation of the system itself, in view of the possibility of advanced technological processes, industrial inventions and scientific discoveries.

This prediction is therefore of extreme importance for the engineer who, without guessing, must attempt to formulate predictions based on credible hypotheses and the instruments at his disposal.

CONCLUSION

It is hoped that this discussion has analysed the transportation question in a generalised way and has shown the need for the transportation engineer to look at problems not as a planner, organiser or designer, but always and only as a transportation expert, taking into account all the diverse aspects which any transportation problem may have.

FOOTNOTES

[1]   These two terms, functional and structural, tie together the facets of planning and operation, both of vehicles and fixed structures. In this manner, an interdependence between the two states is necessarily created. The functions carried out depend on the planning done, whereas the project must be developed with full cognisance of the functions which the system must perform.

[2]   The construction of a vehicle is an act of planning which necessarily also contains some concepts of organisation which are related to the service which will be assigned to the vehicle. The design of the vehicle and, therefore, the vehicle itself can undergo some modifications owing to use-related requirements. But the use of a vehicle is also bound by conditions fixed during the design phase, in the sense that two buses, one for urban use and the other for inter-urban routes, differ in the design of certain parts, namely, the motor, transmission, body, etc. In the territorial planning of transportation, the two aspects, planning and organisation, are almost equal, even if the transportation plan, being limited to the basic project, would appear to exclude this aspect.

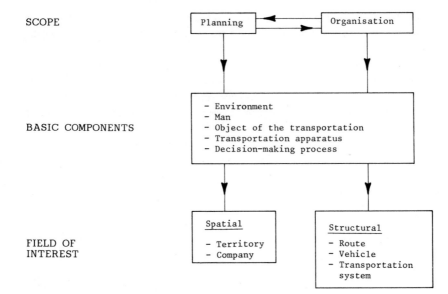

SCOPE

BASIC COMPONENTS

FIELD OF
INTEREST

Figure 1

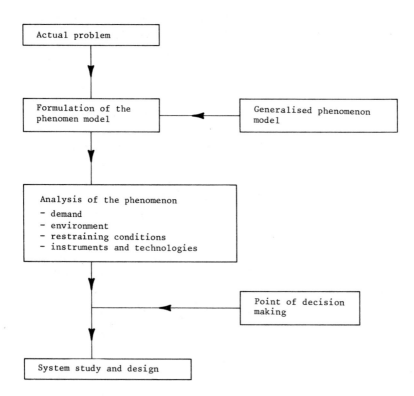

A simplified transportation planning process for South Africa

19

Peter R. Stopher and Chester G. Wilmot   USA/South Africa

SYNOPSIS

With the passing of the Urban Transport Act in 1977, it became mandatory in South Africa for metropolitan areas to carry out comprehensive urban transport planning to qualify for government funds for transport investment and subsidies. As a result, it has become necessary to develop a set of guidelines for urban transport planning within the metropolitan areas of South Africa. This paper describes the process that has been recommended by the National Institute for Transport and Road Research.

The recommended process can be split into two principal components, long-term planning and short-term planning. The long-term planning element involves the use of multiple futures and simplified procedures for travel demand estimation (sketch planning). These should permit the examination of a number of alternative transport plans and strategies for each of the alternative futures. An important aspect discussed in the paper is the fact that the long-term planning effort should take a much smaller component of time and money in the transport planning process than has been customary in the past, and that the procedure should use relatively little data. The recommended forecasting procedures rely rather heavily on synthetic models, instead of on models calibrated on current data. The short-term planning element uses a number of recently developed techniques for corridor analysis, sub-area focusing and localised planning. It is envisaged that intensive data collection will take place only in the short-term planning activity, and that the data collection involved will be focused on the schemes under examination. As a result, the process demands a continuing data collection activity, but one which in any year is at a relatively low level. Over a period of several years, a fairly comprehensive data base will be established.

The planning process is goal-oriented and involves systematic approval by the authorities at various stages in the process. Implementation is tied to the approved plan. It is also designed as a continuing process, wherein the long-term plans are reviewed and revised at approximately five-year intervals, and where the short-term planning is a continuing activity. Within this continuing process, the monitoring of implemented schemes becomes a component element, contributing further to the development of tools and techniques for the estimation process, while helping to build a richer data base. Major emphasis throughout the proposed process is on movement away from large computers and towards procedures that can be used rapidly and without sophisticated facilities.

INTRODUCTION

Recent legislation in South Africa has established administrative machinery and a process whereby apportioned central government funds may be used to subsidise transport costs in urban areas.[1] One requirement of such subsidies is that the supported scheme must be part of an approved transport plan. The current emphasis in the country is consequently on preparing metropolitan transport plans that will satisfy this requirement. In order to establish a transport planning process that will rationalise the activity in this field, the Department of Transport requested the National Institute for Transport and Road Research (NITRR) of the Council for Scientific and Industrial Research (CSIR) to prepared guidelines on urban transport planning. The research was undertaken by the Transport Planning Group. This paper is based on the work of this team, and the guidelines have been submitted to the Department of Transport for consideration.[2]

THE NATURE OF THE GUIDELINES

A central issue in the preparation of the guidelines was the level of specification at which they would be formulated. Detailed specification would provide greater opportunity for guidance to practitioners, comparison between studies, the standardisation of procedures and data and control in the execution of studies. On the other hand, a low level of specification would allow innovative development, a dynamic self-updating process and flexibility in execution. An attempt to combine the desirable features of both these extremes was made by proposing a mandatory framework for conducting urban transport planning, followed by non-obligatory suggestions on the execution of individual aspects identified within the compulsory outline. The fixed skeletal framework would establish certain features for the process, while freedom within its constraints would possibly allow the benefits of a loosely specified process as well.

THE RECOMMENDED COMPULSORY FRAMEWORK

It was recommended that the urban transport planning process be conducted in the manner depicted in Figure 1. The study proposal would contain at least the following essential information: a description of the manner of execution of the study, an explicit statement of what the study aimed to achieve (i.e. the goals) and its expected cost.

The long-term plan was designed to be much less specific in its description of transport solutions for twenty years ahead than has been the case in many long-term transport plans of the past. It was to recognise the uncertainty of the future and that coarse estimates of future transport needs would therefore be adequate. A typical long-term plan would consist of policies and strategies and an outline of long-term infrastructure needs. The policies would state the direction in which transport development should move in the future, while strategies would identify times and coordinated courses of action at a coarse level over the twenty-year period. Infrastructure needs would show facility type and route location, but usually not facility size other than rough estimates of traffic flows.

In contrast to the long-term plan, the short-term plan would need to be sufficiently specific to provide a five-year programme consisting of five individual annual elements. Based on an intestive study of the existing short-term situation and with the long-term plan as a background, the short-term plan would be compiled on relatively accurate short-term estimates, while remaining consistent with the direction of estimated long-term developments. Short-term planning would be done essentially at the scheme level, where alternative schemes would be designed to the basic design level, evaluated in order of priority and scheduled into five individual annual elements. Those schemes not securing a place in the five-year programme would be placed in reserve for possible later use. Before implementation of each annual element, each scheme featuring in that programme would be designed to detailed-design level.

Annual submissions would be made to the authorities as applications for subsidy on activities planned for the coming year. These would contain clear descriptions of what was planned for the coming year as well as the estimated cost of the activities involved. Implementation would be monitored in order to record progress, cost and performance of the schemes. As individual elements of the five-year programme were implemented, new and existing schemes from the reserve list would be used in maintaining a five-year programme or in amending the approved short-term plan in the light of new developments or for information derived from the monitoring activity.

Establishing this compulsory framework for urban transport planning would ensure that each transport planning activity would contain the following features: a continuing planning process; goal-orientated planning; on-going approval by authorities during execution of the study; long and short-term planning yielding the levels of detail adequate for each one's role in directing implementation and anticipatory action; implementation tied to the approved plan; unforeseen developments effecting change in an approved plan; a five-year plan always being available for future planning and budgeting; and monitoring of actual achievement.

SUGGESTIONS FOR CONDUCTING THE PLANNING PROCESS
The compulsory framework outlined above did not specify in any way the procedures to be used in the execution of the activities. In order to provide some guidance in this direction, while yet retaining this aspect of the process entirely at the discretion of the practitioners conducting each study, suggestions are made on how individual aspects of the study could be conducted. Brief outlines of the suggestions on preparing the study proposal, a long-term plan and a short-term plan are given below.

Suggestions on preparing a study proposal. It should be noted that South Africa does not currently use or envisage a competitive bidding process, such as that used in the US. Therefore, the study proposal serves the function of informing the local authorities and the Department of Transport what will be done. It is also envisaged as a mechanism for assisting in the control of a study by the authorities. It was suggested that the study proposal contain a fairly wide spectrum of information which would both enhance the presentation of what was envisaged and clarify certain issues at the outset of the study. It was suggested that a study proposal consist of the following elements:
□ Site plans. These were considered necessary to orientate the reader and describe some of the main features of the study area.
□ A statement of planning goals. Goals were defined as "idealised states towards which a plan would be expected to move - the desired eventual end-states of a planning process". It was suggested that the goals should be established by community representatives by means of questionnaires analysed by psychometric scaling techniques.

□ Main parameters for scenarios. As described later, it was suggested that future conditions should be predicted by means of scenario forecasting.[3] These scenarios would be established by assuming future states of certain basic parameters which would be largely influential in dictating the scenarios. Various scenarios would be developed to portray a reasonable spectrum of possible future states as influenced by the probable range of these basic parameters. In order to allow the authorities the opportunity of viewing these parameters and their possible ranges before progressing with their development, it was suggested that they be described in the study proposal.

□ Study outline. The study outline would describe the execution of the study. Besides the overall description of the process to be followed, an assessment of the duration and schedule of the main activities and a clear exposition of the procedures and techniques to be used would be included. Attention would be given to the appropriateness of the methods employed to achieve the study goals.

□ A public participation programme. It was suggested that a public participation programme as an integral and scheduled activity in the planning process would contribute to the effectiveness and the economic application of this activity. Public participation was seen not only as active participation by the general public in the planning process, but also as participation by the community through elected officials, organisations and surveys, and the dissemination of relevant information to the community. Suggestions were made concerning the compilation of a possible low-cost public participation programme.

□ A description of the organisational structure of the study. The purpose of this activity was to establish the function and responsibility of each body involved in the study. This would allow for the early identification of any weaknesses within the organisational structure of the administration and in the execution of the study itself, and would enable each body to appreciate the role it was expected to play. A description of the manner of interaction between individual parties would be included.

□ A financial statement. The financial statement would indicate funds that the metropolitan authority would have available to implement any schemes recommended in the transport plan. While a probable range of funds is all the authorities would be prepared to volunteer, the exercise is nevertheless regarded as essential in the structuring of alternative solutions.

□ A study budget. The final suggested element of the study was a cost estimate for the execution of the study itself. It was suggested that the budget be compiled by individual task in the planning process, thus providing a valuable reference for controlling expenditure during the execution of the study.

Suggestions on preparing a long-term plan. It was suggested that a long-term plan be conducted by the execution of the activities represented in Figure 2. The four activities at the top of the diagram could be commenced simultaneously.

First, it was proposed that land-use planners prepare a full set of development patterns possible for the area and discuss these with the authorities affected. Study of the broad implications of each, together with the requirements of the authorities, would result in the selection of no more than three development patterns for further investigation.

Secondly, it was suggested that teams of experts, using the basic parameters identified in the study proposal, compile scenarios of the future for each combination of the parameters considered. These scenarios would give particular attention to aspects with a bearing on land use and transportation. The proposals for incorporating a number of scenarios in the transport planning process have been developed by Royce.[4] Unlikely or similar scenarios would be discarded and, considering the individual likelihood of the remaining scenarios, together with the consistency of the transport implications among them and the possible effect of the "worst" scenario, a single scenario would be selected as the "design case". The design case would be a likely scenario (or a combination of similar scenarios which are jointly likely) and one generating plans on which the adverse affect of the realisation of other scenarios would be as small as possible. The design case scenario would be the main one which the long-term plan would aim to satisfy.

Thirdly, it was suggested that more specific objectives for the long-term be developed on the basis of the planning goals established in the study proposal. The preparation of objectives could involve the assistance of representatives of the community and the relative importance of individual goals established by means of psychometric scaling procedures.

Fourthly, it was proposed that an inventory of existing physical features, facilities and their level of utilisation be conducted. Besides providing a picture of the existing situation, such an inventory would identify current deficiencies which could be remedied by either long-term action or immediate action within the short-term plan. Suggestions were made on what data should be collected and the format in which they should be presented.

The information gathered during these four activities would be used in developing a number of internally consistent alternative transport and land-use plans on the basis of personal judgment. The plans so developed would be directed essentially at the design case scenario, but their suitability to other feasible scenarios would be an essential element of the evaluation of the plans. The plans should strive for flexibility (i.e. the ability to accommodate change) and robustness (i.e. the potential to satisfy a wide range of futures), both of which can be assessed through the scenario approach.

To determine the basic travel demand that each alternative plan would generate, it was suggested that sketch-plan tools be used.[5] These are relatively simple, cheap and rapid demand estimation techniques, providing coarse estimates of movement values at the corridor level. Although further development would be required, it was suggested that sketch-planning use existing data as much as possible, consider only the work trip, use a modified four-step procedure and operate on between ten and forty analysis zones. Suggestions of data required include population, employment and car ownerhsip or income on a zonal basis, together with some form of generalised cost of travel. Trip generation would be established by deflating the typical two trips a day for each worker by taking into account the percentage attendance, vacation period, percentage walk trips, etc. Special generators would receive individual attention. The gravity model or growth factor methods were suggested for trip distribution and all-or-nothing assignment for the trip assignment phase. Taking into consideration the design-case scenario, the land-use plan in each case, and the basic travel demand evolving, a modal-split policy would be established and applied in the trip-assignment phase. Adjustments to plans would be made in the light of the policy if necessary. Facility needs would be determined through nomographs relating traffic volumes at various service levels to sizes and types of facilities. A complete set of such nomographs is yet to be developed.

The alternative plans would be evaluated in terms of their ability to satisfy the scenarios (primarily the design case, but also each other possible scenario), their economic performance and their influence on less quantifiable factors such as environment, energy and the welfare of the people. Evaluation of the fit with scenarios and the less quantifiable factors would be done by a panel of experts using their professional judgment. Quantitative analysis would be conducted with standard rates and costs for facility requirements determined from the demand estimation. Selection of the long-term plan was proposed as being dependent on the evaluation, the flexibility and robustness of the plan (determined subjectively) and the likelihood of its implementation. This likelihood could be influenced by factors such as finance, politics and opposition by pressure groups.

It was suggested that the long-term plan submitted for approval should indicate what new infrastructure was proposed and include a schedule indicating when the individual facilities need be introduced; what new schemes, not involving new infrastructure, were proposed and include a schedule indicating when these should be implemented; as well as a statement on what policies should be adopted; the chosen modal split; a cost estimate of the proposals in the plan; and a statement of the cost of conducting the study up to the present stage.

Suggestions on preparing a short-term plan. It was suggested that the short-term planning activity be conducted in the manner represented in Figure 3. Short-term objectives would be developed from the goals established in the study proposal in the same way as long-term objectives were established in long-term planning.

In addition to the long-term plan and the information obtained in the activity describing the existing situation and its deficiencies, an activity involving short-term predictions of pending trouble spots was recommended as an input for the development of alternative schemes. Prediction periods of five to ten years would be typical and conventional methods of prediction would be employed. Personal judgment would be used in identifying areas requiring this attention.

A list of potential schemes would be compiled which would attempt to satisfy the needs identified in the previous activities. The schemes would not necessarily need to be completed within the five-year programme and some might, clearly, become due for implementation only in a few years. The development of potential schemes would involve the public through questionnaires, the press, drop-in centres, a telephone enquiry service, and radio and television discussions. Such involvement would enhance a sense of commitment and support for the schemes arising from the planning activity. Liaison with the authorities was also suggested.

In order to design the above schemes to basic design level, it was suggested that various analysis procedures, suitable at the project level, be used, such as individual choice models[6] [7] and micro-assignment.[8] Guidance was given on their application. Special data collection would probably be necessary and suggestions were made on data-collection methods and the potential for the standardisation of data with a view to building up a valuable data base. Models could be calibrated on the local data or, where possible, from other areas. The procedures used would be as rapid and policy-responsive as possible, in order to allow various alternative schemes to be investigated and the consequences of their implementation determined. It was suggested that in order to achieve simplicity and rapid response, care should be taken to identify the appropriate market segments for a particular policy or scheme; that if possible travel behaviour changes should be defined as narrowly as possible; and that the geographic extent of travel changes should be considered only over a limited area (e.g. small geographic areas or corridors).

For the evaluation of the above schemes, the utility analysis technique[9] supplemented by an economic analysis[10] was suggested. The utility analysis procedure can handle both quantifiable and non-quantifiable factors, and it uses the criteria evolving from the objectives identified earlier in estimating the performance of individual schemes. The procedure is executed by obtaining importance weightings for the identified criteria from a panel and using these to describe the degree of goal attainment and percentage improvement within the current situation. For economic analysis of the costable factors involved, benefit-cost ratio or net present value was suggested for application.

The results of this analysis would be used by the decision-makers (the authorities) in selecting and ascribing priorities to schemes, bearing in mind the financial constraints identified in the study proposal. Five annual programmes would be prepared and schemes not included within these programmes would be placed on the reserve list.

It was suggested that the short-term plan finally submitted to the authorities contain a description of the five annual elements making up the plan; a cost estimate for each element; a description of the schemes placed on the reserve list; a description of how the study was executed, the assumptions that were made and some of the interesting results produced; and a statement of the actual costs of the study.

CONCLUSION

The planning process outlined above is considered to be less labour-intensive yet more effective than the single long-term planning effort conducted in many transportation studies of the past. The features of a more detailed study proposal, scenario forecasting, sketch-planning, the use of individual choice models and evaluation by means of the utility analysis technique are all considered to be improvements on the conventional process which will be perfected with application. Application of conventional procedures is not ruled out. Indeed, they may be applied within the recommended compulsory framework of planning described above. However, the suggestions on conducting the planning process seem to warrant application in order to attain the improvements which they promise to provide.

Another innovation of this procedure is the financial statement of the proposal which is intended as a criterion of evaluation of both the long-term and the short-term proposal. The notion is that the provision of these figures and their use in evaluation should ensure that the strategies and projects proposed by the study are financially reasonable. This has been one of the most serious shortcomings of past transportation studies and has contributed to undermining the value of such studies to authorities throughout the world.

REFERENCES

1   Republic of South Africa, <u>Urban Transport Act, 1977</u>. Government Gazette, 15 June 1977.
2   NITRR, <u>Draft recommended guidelines on urban transport planning</u> (1978), CSIR.
3   Behrens, J., McFeatters, A., and Smith, D., <u>Proceedings: Year 2000 Alternative Transportation Futures Conference,</u> Chicago Area Transportation Study, 1976.
4   Royce, N.J., <u>Scenarios and long-term planning in transportation</u> (1978). NITRR Technical Report RT/3/78, CSIR.
5   Comsis Corporation, <u>Final Report on travel estimation procedures for quick response to urban policy issues</u> (1977), NCHRP Project 8-12 and 8-12A, Transportation Research Board.
6   Domencich, T., and McFadden, D., <u>Urban travel demand</u> (1975), North Holland Publishing Co.
7   Stopher, P.R., and Meyburg, A.H., <u>Urban transportation modelling and planning</u> (1975), Lexington Books, D C Heath & Co., Chapter 16.
8   Creighton-Hamburg, Inc. <u>Micro-assignment</u> (1969), Final Report, Vol 1. US Department of Transportation, Federal Highway Administration.
9   Zangemeister, C., <u>Nutzwertanalyse in der Systemtechnik</u> (1973), Wittemannsche Buchhanlung München, 3 Auflage.
10 Baxa, J.V., and Van Zyl, N.J., <u>NITRR Draft recommended guidelines on urban transport planning: Evaluation</u> (1977), CSIR.

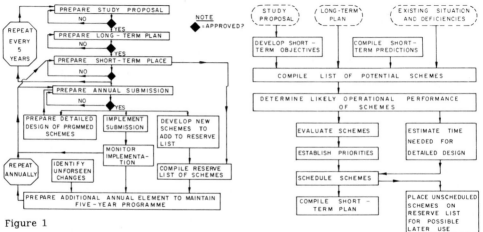

Figure 1

Figure 3

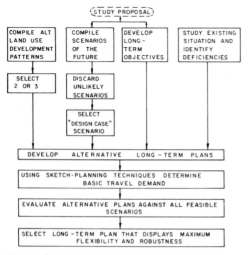

Figure 2

Figure 1. The proposed transport planning process

Figure 2. The suggested process for developing a long-term plan

Figure 3. The suggested process for developing a short-term plan

PART III: REVIEW PAPERS

Transportation engineering, urban development and environmental planning 20

Hans-Georg Retzko and others Germany

SYNOPSIS

The paper offers a general report on the seventeen papers submitted under Theme 1 of the Tel Aviv conference, eleven of which are included in the two volumes of the printed record.

INTRODUCTION

Traffic and transport are as old as mankind, since they mean the movement of people, goods, communications and energy from one place to another. Traffic and transport can be an end in themselves or the means to an end. If they are an end in themselves, they derive from the fundamental need of man to move. As the means to an end, traffic and transport develop from the nature, scale and assignment of land use, so that different kinds of land use can function in their interdependence by surmounting the space between them. Therefore, the traffic and transport problems of cities and regions mostly result from the planning of land use which does not correspond to the specific traffic and transport needs.

From early times, traffic and transport systems have been used for the extension and development of the space in which a human being acts. Therefore, they accomplish essential functions in the development of the environment. At places where transport has been developed to a high degree, economic activities as well as human wealth have grown. The widespread use of the private motor vehicle and the extension of different kinds of public transport on land, on water and in the air have given nearly everybody the utmost individual freedom and mobility. We must not forget these and other positive effects on human life when the adverse effects of transport are discussed.

Traffic and transport planning itself should also be conscious of these relations. Those who plan traffic and transport must not so much cure symptoms as consider the primary causes of traffic generation and its relevant factors.

The first theme of the Tel Aviv conference was intended to deal with the complex and comprehensive problem of the interdependence of transportation engineering, urban development and environmental planning. The sub-themes were:

☐ Land-use and other non-transportation solutions to transportation problems (i.e. reducing the amount of travel by means of urban planning concepts).

☐ Economic growth and the provision of mobility for all population groups.

☐ Interaction between traffic facilities and urban development.

☐ The integration of public transportation with urban planning (bus-oriented town patterns).

☐ Minimising traffic noise, air pollution and visual obstruction through mode of travel and urban planning.
☐ Transportation safety in towns.

STATISTICAL REVIEW OF THE PAPERS
Seventeen papers were announced for the first theme. The response can be interpreted as a reflection of the importance of various questions in the field of traffic and town planning. Problems of noise and other environmental impacts are discussed in five of the papers while five deal with different forms of interdependence between traffic facilities and urban development. Other subjects seem to be of lesser interest.

SURVEY OF THE PAPERS
A general report should reflect the essential points of the individual papers and highlight conflicting or new aspects of the problems discussed. Unfortunately, the number of papers delivered on the various sub-themes was very small. A comprehensive comparative evaluation of the papers was therefore impossible. Instead, a brief survey of each of the papers will be presented in the order of the themes, and this will be followed by conclusions, subjects proposed for discussion, and a final statement.

Land-use and other non-transportation solutions to transportation problems
☐ Eliezer Brutzkus attempts to explain the interdependence of transportation and urbanisation patterns from the point of view of the historical development of urban life. Owing to the existing transportation facilities and strategic needs (e.g. defence walls) the growth of towns was limited for centuries while the population density remained about constant at 80 m² per inhabitant. In the past century this ratio has increased to 200 m² per inhabitant and the change in population density continues in the same way as private motorisation has not yet reached saturation point.

The author distinguishes between two fundamentally different city models which represent two stages. The first is a model of a city with a high population density and a large CBD dominated by public transportation, and the second is a model of a city with a low population density where the CBD has nearly disappeared and private cars dominate. In the second model, land consumption rises to 600 and 700 m² per resident. Most cities are sitll a mixture of the two models, however. In the long term, motorisation is a self-defeating process because congestion in CBDs and the daily commuting distances will be so great that time gains from the change to private cars will be neutralised even for the car owner himself and certainly for society as a whole. The only advantage will be the spaciousness of settlement conditions which is, however, a matter of personal taste and preferences. At present, one tries to achieve a stable symbiosis between the existing compact cities and low-density suburbia. This means restrictions on the use of cars within the centres of urban regions and further development of public transport. But even very ambitious attempts to achieve that aim, such as BART in San Francisco or the mass rapid transit system of Stockholm, were only partly successful. Today's state of a mixture between the two models is described as sub-optimal from almost every point of view, so that the trend to low-density areas cannot be changed. If car ownership continues to rise, the transition from the first model of urban development through a sub-optimal (present) mixture stage to the second model is unavoidable. Many densely populated countries cannot afford low-density urbanised regions because of high costs and scarce agricultural land. Therefore, a new policy of protecting open countryside and preserving compact cities must be accompanied by an equally strong transportation policy discouraging private motorisation and ensuring the dominant role of public transportation.

To improve our understanding of transportation problems, we have to go back to the conditions which generate a given pattern in which people go about their daily lives.

□ Velibor Vidakovic reports on travel activity research by analysing the transportation process in its close relationship to urban, socio-economic, temporal and land-use orders. The method applied in this study is a disaggregate observation, analysis and modelling of individuals' travel behaviour under the choice constraint conditions created by social, temporal and spatial variables. An individual sequence of activities and trips during the shortest cycle of twenty-four hours is considered as a basic unit of transportation analysis. An explorative study carried out in Amsterdam shows a variety of patterns and indicates the relevance of the chosen approach for the land-use and transportation research. The conclusions note that redefined mobility and accessibility concepts should embody the activity variables. For example, the objectives in land-use and transportation planning should be primarily concerned with reducing the amount of travel for obligatory activities so as to stimulate the development of non-obligatory, non-home activities. It is suggested that to the evaluation of alternative land use/transportation a new criterion should be applied - transitiveness, i.e. the degree to which the systems stimulate the transition from one activity to another.

Economic growth and the provision of mobility for all population groups

□ K.J. Button and A.D. Pearman report on problems in forecasting car ownership by land-use patterns and the impact that changes in these patterns will have on future ownership levels. The authors do not consider the extrapolation approach appropriate for urban planning exercises because data over a long period are needed for calibration, and these are not always available for specific urban planning problems. Cross-sectional techniques were therefore employed in which car ownership is directly related to a set of causal variables. The empirical investigation is based on two major assumptions: first, that household-based causal models provide the appropriate framework for a car-ownership model; secondly, that essentially policy-sensitive variables have to be incorporated explicitly in the model. Several types of policy-sensitive variables were tested and the evidence of their influence could be proved. Also it is concluded that significantly different patterns of behaviour are exhibited by small and poor households so that possibly no single forecasting framework can be derived for the entire population.

□ L.R. Soares describes tendencies of urban development in terms of the Rio de Janeiro situation, and their consequences for the volume of traffic. The fundamental changes first observed as characteristics of American cities now occur elsewhere on the same scale. For example, commuter streams between suburban areas and the CBD arise, causing far-reaching changes in former trip models, especially in home-work-home relations. Urgent problems of traffic congestion result from this and their worst effects are felt not within the CBD but 3-6 km around it. At these distances, different traffic flows have to converge and mix. In peak hours most private vehicles transport only the drivers, and most taxis only a single passenger. The implementation of private car-pools is proposed as a counter-measure: several commuters working in neighbouring places and living close to one another arrange to use only one car for their daily trip to work. In Rio de Janeiro there is no effective system of mass transit. After a seven-year experiment with trolley-buses to replace the former light rail system had failed, a large network of diesel buses was established. The possible benefit of a subway still under construction cannot be estimated precisely until it goes in operation. So it must be evaluated in comparison with other towns where similar systems operate. In order to find the required number of radial lines to fit optimally to the needs of transport, the author proposes to use a simple straight line formula:

$$n = k \cdot R$$

where k is a local factor and R is the radius length in kilometres of concentric circles around the CBD. Such a subway network with a bus system of feeder lines could offer a good standard of public transport service.

Interaction between traffic facilities and urban development
□ Model calculation is usually applied as the most suitable procedure in analysing travel and spatial interaction. Katsunao Kondo and Tsuna Sasaki report on an entropy-maximising model of current type which is estimated to offer a better understanding of the solution structure of conventional entropy models. Up to now many publications have designated the relations between models of different approaches as gravity type, entropy model, utility or other behavioural theory. Indeed, it is very interesting to derive (originally empirical) gravity pattern theoretically in order to permit comparisons with growth factor techniques. For this purpose, it is necesssary to introduce one additional piece of information into the process of trip assignment between given origins and destinations.

The practical procedure is based on the condition that the origin-destination pattern of the horizon year should be most similar to that of the base year. This condition includes various optimisation problems, for example the minimisation of the residual sum of squares or of chi squares or the maximisation of joint probability of occurrence. Instead, the authors introduce the condition that a priori probability of the $(i,j)_{th}$ cell in the horizon year origin-destination table must be equal to the unit origin-destination table of the base year. Mathematical treatment of this approach leads to solutions directly comparable to those of the Detroit model growth factor method. The only difference is in the technique for the convergence criterion. Friction factors, corresponding to local conditions, can be included as further constraints.

In the last section of the paper a model-based method is described to estimate the actual or future distribution of urban activities in a city, particularly the number of households. The main equation to locate the population in each zone of the study area is directly derived from the entropy-maximising trip-distribution model of the gravity type. On the assumption that trip distribution forms a gravity pattern, the most probable pattern of household locations results from this model equation.

□ Living conditions in urban areas, as they are affected by traffic impacts, are popular criteria all over the world. Harold Marks presents a study summarising the experiences of numerous American cities of various devices and techniques to protect residential neighbourhoods against extraneous through traffic. Originally, the amount of neighbourhood-generated traffic only depends on the number of dwelling units, car ownership and other structure characteristics. Nevertheless, it can be seen that the number of vehicles using a residential street is proportional to its continuity and directness. These criteria make it profitable for through traffic to leave congested arterials and use local streets. So all programmes for decreasing environmental deterioration by traffic should start with the examination and improvement of alternative routes. The directly local devices result from three potential approaches: protection, amelioration and compensation.

Practical counter-measures include traffic controls, traffic deterrents and traffic diverters which aim at restricting, impeding or prohibiting traffic flow within residential neighbourhoods. The effectiveness of the chosen devices varies according to the degree of restriction imposed. Minor traffic controls (such as stop signs, traffic circles and turn prohibitions) often prove ineffective, at times even counter-productive. However, higher levels of restriction, as realised by physical deterrents and diverters, impose greater travel circuity as well as inconvenience and arouse greater opposition. Intermediate restrictions are frequently applied as a compromise, providing substantial relief with minor inconvenience. Partial street closures may effect reductions in traffic flow of up to 50 per cent and reduce accident rates up to 90 per cent. Local success will always depend on the balance of various measure in the programme applied.

□ E.A. Rose and P. Truelove give some policy considerations on the various ways of decision making on transport investments. Their paper is based on comprehensive comparisons of recent rapid transit projects in Britain and France. Whereas the English tradition of the non-specialist administrator contrasts strongly with the engineering-based technocracy of the French administration, many similar factors are found in both countries and they influence the process of deciding in the same way. Consequently, even some results are comparable. It is interesting that technocratically generated proposals at various government levels are initial starting points but never the result of any expert study or mathematical transport models. As to the chosen technologies, an examination of various transport studies proves that the conclusions favoured correspond to those material systems most familiar to the body working out the special plan. Compatibility with established and new lines is of additional importance. Another important influence on the decisions is the expected development potential of a new or improved transit system. Growth and development rather than the renewal or revitalisation of urban areas is considered and aimed at. Cost-effectiveness relations may not be regarded as the only or decisive criteria, especially as widespread network proposals must often be replaced by combinations of several smaller local-bound measures demanding less investment.

□ Michael C. Poulton reports on a general model for the replanning of existing residential access and secondary arterial street systems in order to improve systematically the residential environment at minimum cost to road users. The model postulates a uniform rectangular grid street network servicing a bounded residential area traversed by primary arterials. The incorporation of boundaries and primary arterials into the model is important because the model is applied to Vancouver, which is bounded by water on three sides while its primary circulation system consists almost entirely of at-grade arterials.

An idealised model is used that states total travel time as a function of the road pattern, the traffic control system and various travel characertistics. The results obtained from the model suggest that the residential environment of Vancouver could be improved and travel times reduced simultaneously.

□ Peretz Wittman, on his experience as a member of the planning department of the London Borough of Lambeth, reports on planning problems in the Brixton town centre, one of the 55 town centres in the inner and outer suburbs of London. In the first part of the paper the statutory background and changes in planning Acts and guidelines are described. He shows how much planning practice to structure actual local development depends on and can be delayed or even blocked for want of a comprehensive master or general plan. The immobility of the upper planning authorities leads to uncoordinated planning or even no planned development. In the second part, a chronological survey of proposals to solve the urban and traffic problems of Brixton is given. One after the other, these proposals failed or were cancelled because they favoured development or replacement instead of remedial measures to improve the organically grown structures. The conclusions finally drawn indicate that "all or nothing" approaches have little chance of being realised, because these concepts are unable to fit in with local conditions. The huge investments required and the changing public and political opinion are other reasons for this failure. The actual task is to recast planning approaches in terms that are more sensitive and responsive to accomplish the original objective of renewal.

The integration of public transportation with urban planning
(bus-oriented town-patterns)
No papers were submitted for this sub-theme.

Minimising traffic noise, air pollution and visual obstruction
through mode of travel and urban planning
□ Most environmental problems of larger cities are caused by commuter traffic because work places and dwellings are separated by great distances. Karl Krell reports that such a separation is not absolutely necessary because the disadvantages of work places for their neighbourhood in terms of noise and air pollution have been significantly reduced. So work places of the third sector (such as banks, insurance companies and offices) can be relocated in dwelling places, or vice versa, in order to reduce the volume of commuter traffic. In the existing historical towns this can probably not be achieved within a short time but it may be a concept for developing areas or those to be redeveloped. In these cases, work places should be constructed in such a way that they protect residential areas against traffic noise and air pollution. This can be achieved in a ring-shaped buffer zone which forms a unit for about 5000 inhabitants.

□ In the urban centres of today the conflict between people and vehicles is mostly solved by the decisions of specialists and authorities relying on traffic manuals or books of warrants which widely consider motor traffic as of paramount importance. It has been suggested that the warrants regulating pedestrian/vehicle interactions are in some cases unsatisfactory. Even in the recently planned "new towns" the rights of pedestrians to safety and convenience could not be realised as well as in old cities, such as Cordoba or Jerusalem. Colin J. Taylor proposes to use the wide range between the two extremes of pedestrian-barred freeways and motor vehicle-barred precincts by mixing the two modes of traffic.

Two measures should be generally adopted to achieve the aim of balanced coexistence between pedestrian and vehicle traffic: all pedestrian crossings of the zebra-type should be raised at least 10 cm above the road surface and be of a width commensurate with pedestrian flow; and all roads at the site of and approaching pedestrian crossings should be physically divided into single lanes by narrow refuge strips at least wide enough for a pedestrian to stand in safety. The narrowing of lanes in the vicinity of pedestrian crossings would physically constrain the traffic by slowing it down.

A list of new factors should also be included in any future warrants governing pedestrian/vehicle interactions to give greater consideration to the characteristics of pedestrian traffic and to back up its physically weak position against the automobile.

□ Of all possible measures against traffic noise, the most effective and inexpensive is the re-arrangement of the deployment of land use. The contribution by Valerie Brachya and others describes a comprehensive approach to establish a traffic noise model in order to forecast traffic noise. If noise disturbances are known before the construction of roads and dwellings many options are open for appropriate countermeasures.

The existing noise model is applied to Tochnit Lammed, a residential area north of Tel Aviv, where it is found that the expected noise level will reach the 65-70 dB(A) category. As noise standards adopted in the United States and several European countries are not appropriate for Israeli conditions, some standards are developed for the specific needs of Tochnit Lammed. According to these standards, the noise level must be reduced to the 60-65 dB(A) category.

As the area is crossed by major inter-urban traffic arterials, traffic management and route location together with sound barriers were the main factors considered to reduce the noise level. Other counter-measures are also discussed. Further studies will complete the guidelines for the case study so that they can be adapted to planning residential areas elsewhere.

□ G.E. Frangos presents the basic parts of a traffic impact study in Anne Arundel County, Maryland, USA. These parts are the legal basis of traffic flow analysis, a negative declaration form for processing minor subdivisions, a brief description of the project and the coordination of trip generation and trip distribution procedures, as well as peak-hour factoring. Guidelines for the acceptable conduct of the study are presented in an appendix of the report. General criteria, such as traffic generation, peak-hour percentage, peak hours, trip distribution, traffic split, trip assignment, critical lane analysis, traffic data, adequate accommodation of traffic, and final evaluation of adequacy of access, are outlined. In addition, the critical lane volume technique is described for simple two-phase or unsignalised intersections. Another appendix concerns a research study to obtain data on vehicle trip generation for various types of land use.

Transportation safety in towns

□ David C. Andreassend describes a simple technique to analyse traffic accidents in order to discover the shortcomings of present street patterns. In 1968 an accident record system was introduced in certain Australian cities, based on an accident location system to which an accident classification system was added. The results of these studies were registered in collision diagrams. By these means high-accident frequency locations could be identified, as well as typical accident road user movements. Based on the data of 20,000 accidents in the Melbourne metropolitan area reported in 1968, a highly significant correlation between population density and the number of accident locations per area could be proved. Recent evaluations of analogous studies for the cities of Taipei and Kuala Lumpur and comparisions with the Melbourne results show good conformity for the most frequent accident types and distribution on arterial, sub-arterial and local streets. At least 60 per cent of all accidents occur at intersections and 50 per cent of these accidents are located at scarcely a third of the intersections. This means that most of the total number of intersection accidents occur at only a few intersections. Improvement of these critical intersections can effect a reduction of accidents out of proportion to the number of intersections.

□ A.S. Hakkert and J. Bar-Ziv studied the accident risk to pedestrians crossing the road away from intersections (mid-block crossings). Three types of crossings are dealt with: zebra-crossings, near zebra-crossings unregulated crossings. The three major components involved in accident exposure, risk and accident probability are described by equations and data sets chosen from different countries on the common basis of accidents during a four-year period and average hourly vehicle and pedestrian volumes for correlation and regression analysis. The results of the analysis show that no kind of crossing installation is safer than a zebra-crossing up to a ratio of pedestrians to vehicles $P/V=0.02$. Up to a ratio of $P/V=6$ the area near a zebra crossing is more dangerous than other areas. It is, therefore, worthwhile to install pedestrian barriers in order to reduce the risk. The relationship between vehicle delay, pedestrian delay and accident risk, and the relationship between the number of vehicles stopping before a pedestrian crossing relative to the total traffic flow and accident risk are investigated. It can be demonstrated that the actual risks at each type of location are very sensitive to the ratio of pedestrians to vehicles and that installations perform differently according to the ratio of activities.

Unclassified contributions

□ In his invited keynote paper, Herbert S. Levinson looks at the 21st century American city. In terms of development patterns of the past, he describes the trends of today and the influence they will have on cities, life-style and mobility in future. Based on a historical perspective of technological, social and economical factors, growth trends for the years 1975-2000 are outlined under the conditions of no war, no sustained oil embargoes or major economic recessions. Some views of the 21st century metropolis present various futures of different planning horizons for the years 2000, 2025 and 2050. The metropolis of 2050 will reflect the impacts of new technology as well as public policy. Public transport services, private transport vehicles and pedestrians will each have their own separate rights-of-way. That means full application of the principle of access control.

The new cities of the 21st century will be built in an environment-sensitive and energy-conservant manner. In this way, they will be able to remain the centres of culture and society in the US.

□ J. Dash first describes Israel's planning and building law of 1965. The planning administration consists of three levels: the national planning board, the district planning commissions and local planning commissions. The national planning board has thirty-one members, of whom eleven are representatives of local governments (among them the mayors of the three largest cities in Israel) and ten representatives of public and professional organisations. The national planning board prepares the national outline scheme, approves district outline schemes and advises government in all matters of planning and building in Israel.

The paper describes national outline schemes for national parks, nature reserves and landscape preservation, national road systems, public institutions, tourist facilities and recreation sites, storage and distribution of flood waters, power stations and electric networks, airfields, centres of noxious industries, and fuel service stations. The most important functions of the district planning commissions, which are composed of fifteen members (nine givernment officials, five representatives recommended by local authorities, one expert in planning and building) are the approval of local outline schems or local detailed plans and drawing up district outline schemes. The paper describes regional outline schemes for districts in Jerusalem, Tel Aviv and the south of Israel. The main tasks of local planning commissions include the preparation of local and detailed schemes, reparcellation of lots, and the issuing of building permits. The local planning procedures are described in terms of some examples given for Beersheba, Eilat, Arad, Nazareth and Jerusalem. The old city of Jerusalem and its surroundings is one of the problem planning sites on which opinions are strongly divided. The urban planning unit will re-examine the master plan, ensuring maximum participation by the public and professional bodies in the planning process.

CONCLUSIONS AND SUBJECTS PROPOSED FOR DISCUSSION

In terms of the interdependence of traffic or transport and land use, briefly outlined above, special conditions of adaption have to be taken into account to achieve a satisfying plan which really integrates all aspects. This is the starting point for coordinated or even integrated planning activities of the administration and planning authorities. Comparisons of statutory and organisational backgrounds to planning in several countries show that there are wide structural and substantial differences. However, the distinction of different levels and sectors of planning cannot be regarded as the only common principle. It is a surprising matter of fact that in spite of different local conditions, the aims and the criteria of decision making often coincide and the proposals or results are similar. Furthermore, it is to be noticed that, especially in countries where traffic has increased to the worst levels, planners are criticised by a steadily growing number of popular movements. Many of the actual problems result from the discrepancy between today's traffic volume and the historically developed urban structures. Whatever the aims of the planners, there is always one essential step in the process of planning, no matter what methodology may be chosen, and that is the prognosis. Nobody is able to figure out precisely and with certainty what will happen in the future. (The only prognosis that is certain is that the future development of traffic-relevant factors and traffic itself is uncertain.) Nevertheless, it is usual to derive the probable development from the current status. For example, future town patterns can be outlined on the assumption of undisturbed continuity of current trends. Or the rate of private car ownership can be forecast by means of mathematical models based on statistically proved patterns of travel behaviour. Mostly, planning initiatives have their origins in evident or proved deficiencies. The simple conclusion, that planners must look for defects and needs from which to derive their objectives, can scarcely be denied. The most serious deficiencies are in the field of traffic accidents. It is a reasonable approach to evaluate traffic accidents systematically in order to acquire information on how to improve single intersections or even greater parts of a network. In the same way, observation and analysis of congestion or faults in mass transit systems can stimulate improvement of public transport services.

The increasing public interest in questions of planning, and especially traffic planning, is derived from a need to protect the environment against pollution and negative impacts which are still growing. Undoubtedly this concern is legitimate, but it does not mean that forms of protest which surpass all legal limits can be justified.

What can be stated in general is that worldwide efforts to reduce or even eliminate the conflicts between traffic and environment seem to be of the greatest importance.

One of these efforts concerns the reduction of vehicle traffic within housing areas in order to improve safety for pedestrians and environmental conditions on the whole. We all know the many activities in this field in various countries. Though all these activities may be highly desirable, one must warn against exaggerated measures. As for reducing traffic within housing areas, there is the danger that one may go from one extreme to another. Exaggerated measures to reduce traffic within housing areas can intensify traffic in other areas. Also, it is possible that accessibility for vehicle traffic will be reduced, causing other disadvantages to develop.

This remark may lead us to the problems which pedestrian areas within the central business districts of our cities have created whenever these areas were not planned by interdisciplinary teams. We all know the great advantage and the positive influences on urban life that pedestrian areas have. We need not talk about them. We also know the adverse side-effects of pedestrian areas and we must try to minimise them.

I should like to recommend the following themes for further discussion:

□ Experience in different measures to reduce the adverse effects of vehicle traffic within housing areas.

□ Criteria and principles for the installation of pedestrian zones in central business districts in small, median and big cities.

These very precise and specific items may be used to start our discussion. If time permits, I recommend the discussion of a more complicated and comprehensive subject, such as accident-prevention measures to improve the quality of urban life and land-use planning to minimise travel demands.

FINAL STATEMENT

Theme I of this conference was intended to deal with the complex and comprehensive problem of the interdependence of transportation engineering, urban development and environmental planning. The papers submitted dealt with very different aspects of this theme. The general reporter offered some general statements, and recommended two specific subjects and two more complicated and comprehensive ones for discussion. The very different and specific papers enabled us to gain insights into neighbouring problem fields and to be stimulated by them. We may learn that the same problem may be treated and solved by different means and with different results, according to local conditions. Although they must be considered in rather general terms, I should like to state the following as my conclusions on Theme I.

□ Traffic and transportation planning should always be conscious of the interdependence of land use on one hand and travel demand on the other. (For example, no traffic and transportation-orientated land-use planning may create traffic and transportation problems from which other adverse effect may arise.)

□ By means of new strategies and procedures, traffic and transportation planning must be attuned to the actual changes in population and economy, land-use patterns, transportation policies, and values and attitudes. (For example, short-term flexible planning has become more necessary.)

□ Traffic and transportation planning has to protect the human environment against adverse effects of traffic and transportation. (For example, efforts should be increased to reduce traffic in residential areas, but exaggerated measures should be avoided.)

□ On the other hand, traffic and transportation planning has to serve people's needs and permit the necessary and - from the point of view of society - desirable mobility, by managing the adequate transportation sub-systems as integrated parts of a comprehensive transportation system. (For example, accessibility of areas has to be maintained or even improved by traffic and transportation facilities.)

□ Traffic and transportation planning must not only try to cure symptoms. More creative work is necessary to look into further possibilities and restrictions related to changes in traffic-relevant factors. (For example, factors of land and energy consumption, as well as ecological and social aspects, have to be considered, and new technological and management measures have to be developed in the field of traffic and transportation.)

□ All in all, traffic and transportation planning must be orientated to the goals and objectives defined by social requirements. Within the modern continuous planning process as an integral part of the decision-making process, evaluation goals and objectives has become an essential element.

My personal recommendation to the conference organisers is that the conference, should be repeated after a sufficient lapse of time, and possibly with the subject-matter regionally restricted. Such a conference should deal with fewer subjects. Overlapping should be avoided. The subjects should be chosen from the broad field of traffic and transportation engineering. Some of these subjects, especially those treated controversially at this conference, should be dealt with in in-depth discussions on the basis of case studies. In this way, we will have more concrete and precise exchanges of experience, knowledge and opinion. Then nobody will be disappointed by too general or even superficial statements.

ACKNOWLEDGMENT

I should like to acknowledge the assistance of Friedmann Brühl and Reinhard Forst in the preparation of this general report.

Methods and procedures of transportation planning   21

Paul H. Bendtsen   Denmark

SYNOPSIS

The paper offers a general report on the fifteen papers submitted under Theme 2 of the Tel Aviv conference, nine of which are included in the two volumes of the printed record.

INTRODUCTION

In the discussion of the papers submitted, a division into three sub-themes has been adopted. Papers dealing with more than one of the listed problems may be mentioned more than once in the discussion. The sub-themes are:

□ Simplified planning procedures, including trip generation and the accuracy of transportation models.

□ Economic and environmental considerations and priority schemes, including analyses of intangible factors.

□ Public transport, including school bus operation and pedestrian models.

The first two sub-themes deal with transport in general, but mostly with automobile traffic. Several new ideas are contained in the papers, but the application of the ideas is seldom mentioned in detail. Only Daor's paper deals with trip distribution models. Modal split models are mentioned in Ben-Akiva's paper and assignment problems are mentioned briefly in two reports.

SIMPLIFIED PLANNING PROCEDURES

A general comment is that perhaps not all five papers are so very simplified.

□ Peter R. Stopher and Chester G. Wilmot discuss the evaluation of three different development plans. Different scenarios with varying parameters, population, employment and car ownership are investigated. Only 10 to 40 zones and a simplified four-step model are used. Goals are established by questionnaires and psychometric scaling techniques are used. Prediction of trouble spots is used to produce a list of potential schemes, which are evaluated by utility analyses capable of handling both quantifiable and unquantifiable factors. Weighing up is undertaken by a panel.

□ Moshe Ben-Akiva and others refer to the use of small-scale surveys without loss in policy sensitivity. Use of pocket computers is also mentioned.

□ Arturo D. Abriani suggests that the environmental capacity (see also the second sub-theme) should be investigated against the demand for travel. Assignment is mentioned. The planning problems are limited mainly to the provision of sufficient capacity on distribution roads. The modelling work is thus rather simplified. How future traffic volume is calculated is not mentioned in detail. A similar approach was used in the Edinburgh transport study.

□ Uzi Landau and Yaron Fedorowicz discuss transit in an urban corridor. The model is based on the corridor and the complementary part of the urban area. A sample of 150-300 observations (households) is drawn from the corridor zones. Observations are handled on a disaggregate level. The trips are assigned to the network by an iterative process until equilibrium between assignment and mode split is reached. □ Yacov Zahavi emphasises the model's lack of accurate responsiveness to road pricing and the effect of petrol shortages on travel patterns. The Crystal model developed by TRRL in London could have been mentioned as a model used for predicting the effect of road pricing. The effect of the Singapore CBD experiment, in which a fee was imposed on each car and and taxi with fewer than four persons in the rush hour, was mentioned. An attempt to calculate this effect was not shown.

Trip generation

□ Richard J. Brown (Paper 12) suggests an improvement in category analysis. Instead of the conventional variables, he proposes some new variables: stage in life cycle (young, families, elderly); life-style (income and social group); and transport availability. A new mathematical model for determination of trip rate for each category is proposed. The Yates algorithm is used.

□ Stopher and Wilmot consider only work trips by deflating the two trips per day for each worker. Commuter traffic must then be blown up to total traffic. How that is done is not mentioned.

□ Erella Daor refers to the Greater London study in which income was found to be significantly correlated with both work trips and non-work trips. However, income contribution to the explanation of variation when the effect of the variables was accounted for was less than 1 per cent. The distribution of trip length was also compared.

□ Zahavi mentions that in Kingston-upon-Hull (population 345,000) cars do 6.2 trips per day and in London (population 8.8 million) 3.3 trips per day, and he suggests city size influences the number of trips per car per day. Overgaard from Denmark, however, could not find any relationship between the numbers of trips per car per day and city size in 51 US cities where these figures were investigated. Zahavi in his paper introduces mean travel length and speed in the calculations and there seems to be a relationship between all these variables.

The accuracy of transportation models

□ Brown indicates that the coefficient of variation for individual trips can vary between 100 per cent and 700 per cent. In a study in the UK it was found that the expanded survey data only produced 4 per cent of non-home-based business trips which were recorded on a screen line.

ECONOMIC AND ENVIRONMENTAL CONSIDERATIONS AND PRIORITY SCHEMES

□ Vedia Dökmeci proposes that all city systems (traffic, water, schools, and so on) should be planned together. In each system the total cost of construction and flow should be minimised. The effectiveness of each system is calculated by summing up weighted efficiencies of alternatives in a step-by-step procedure. Further research is needed.

□ H. Ayad and J.C. Oppenlander's paper discusses the determination of levels of service of trips between pairs of zones. Link deficiencies are expressed in vehicles per hour and occur when assigned volumes exceed service volumes. Link deficiencies define where improvements are needed. A simplified proportional assignment technique is introduced and explained in detail. It is determined how a new stretch of road may reduce the number of vehicles running with service level below C. This may be a new planning tool.

□ Karl Ruhm presents a study of a five-year priority scheme for the improvement of the Israeli national road system. Israel is divided in 35 zones and 400 road sections. For each section, capacity is calculated according to HCM, and traffic five years ahead is estimated on future population and car ownership in each zone. A formula for the relationship between traffic and the two parameters is given.

□ Shogo Kawakami states that in a transportation system, which is the sum of all people's dissatisfactions, the minimum dissatisfaction is the best (including cost).

□ Stopher and Wilmot suggest that a financial statement should be given for each proposal. This was not done in many previous studies and has been one of the most serious shortcomings of past transportation studies.

Analyses of intangible factors

□ Domenico di Noto proposes that a list of determinant factors, social and aesthetic, should be prepared. The list should be weighted by the affected citizenry. Ranking calculation should be used rather than counting calculation.

□ Kawakami describes how the percentage of people dissatisfied with each consequence, such as convenience and environment, is determined (questionnaire) and related to distance from street. On a street in Nagoya, the rate of dissatisfaction about environmental effects was 0.5 close to the street and 0.2 at a distance of 210 metres from the street. The Thurston scale was used to determine the relative weights of consequences. The following relative weights were found: air pollution, 1.38; noise, 1.25; danger, 1.0; and encroachment of view, 0.51.

□ Abriani discusses an analysis in which no questionnaires are used. The environmental function is measured by pedestrian delay and noise level. The environment deficiency index is defined as the ratio of traffic volume to environmental capacity, which is the same as maximum acceptable traffic volume. An acceptable level of service for traffic could imply unacceptable social or access functions.

□ Kawakami, Wilmot and Brown mention psychological models and measuring techniques. Brown also suggests that transportation should be taught at universities as a complete subject, and that graduates should be called transportationists.

PUBLIC TRANSPORT

□ Eric L. Bers gives a general description of various types of public transport lines and emphasises that the function of public transport lines is movement and access. The paper distinguishes between three types of public transport lines: systems operating on local streets; systems on partially controlled right-of-way, possible with green signal pre-emption; and systems operating on grade-separated rights-of-way. The evolution of the public transport system in a new suburb is explained, beginning with car-pools and a subscription bus operation. A busway can shape development patterns.

□ Hagi Gil and David Shlay provide an example of the redevelopment of a transit line of the third type mentioned above, i.e. the eighty-mile South Bend Railroad in Chicago. The line was opened in 1901 and the number of passengers per day dropped from 6000 in 1945 to 2000 in 1975. A transportation study suggested nine options, ranging from modernisation of the line to express buses instead of trains. Other options suggested trains on part of the line and express buses on other parts. Modernisation of the railroad was chosen. Patronage of 11,000 passengers per day is expected.

□ Hartmut Keller and Wilfried Müller deal with a special kind of public transport: school bus operation. A mathematical model for determining the optimal route for a school bus was developed. Two objective functions are considered: the first is that total travel time for students should be minimised; and the second is that total travel distance of the bus should be minimised. They show how the two objectives gave different routes. The problem was formulated as a mixed integer problem and a standard program package was used.

□ Brown suggests that the lack of data on pedestrian movements in normal traffic studies is the reason for the inadequate facilities for pedestrians in many CBDs.

CONCLUSIONS

The authors presented many alternative methods and procedures that can be used to collect travel pattern data, to forecast flows and levels of services on transportation facilities, and to evaluate alternative plans and policies. Some of the papers showed how recent advances in techniques can be employed in various ways to improve our analysis capabilities. Intangible factors should be analysed. The percentage of dissatisfied people might be determined by use of questionnaires. The weights of the various consequences (noise, air pollution, etc) should be determined and included in the final evaluation of the plans. Priority schemes should be established.

Further research is particularly needed in the area of land-use modelling. An important criterion is the accuracy of analysis methods and particularly the validity of mathematical models. This issue was addressed by several authors. It was suggested that the validity of old models be tested by comparing their forecasts with traffic counts. Simplified models may not be less accurate than sophisticated ones, and much money can be saved by using simplified models.

Finally, several participants complained about the lack of communication between model developers and model users. The statisticians should perhaps try to express themselves in such a way that they are understood by the practical planners. There is a need for a simplified presentation of sophisticated techniques, expressing the capabilities of new methods in language that can be understood by the users of analytical techniques. The statisticians should not forget to ask the planners what their problems are and what questions they want answered by the models. At a recent meeting in Stockholm on traffic models, it was stated that transportation planners usually fall in two groups: builders of mathematical models and practical planners, and that these groups have difficulty in understanding each other. I hope that this conference contributes to this difficulty being overcome.

Design principles for transportation facilities and operations

22

Sam Cass   Canada

SYNOPSIS

The paper offers a general report on the nine papers submitted under Theme 3 of the Tel Aviv conference, five of which are included in the two volumes of the printed record.

INTRODUCTION

The papers deal with a diversity of design principles ranging from air terminal design to transit vehicle design. The authors are from widely separated geographic locations, which makes this session truly representative internationally. A general report attempts to seek out common elements and to illuminate any new trends in principles or in attitude which may be signalled by the contents of the papers. In this regard, all but one of the papers deal with matters relating to public transportation of one form or another. This in itself is remarkable, since for the past many years a session of this nature would have elicited papers primarily concerned with the design and operation of roads and highways. Moreover, more than half the papers are concerned with design principles and operations which affect the passenger as a pedestrian. Part of any trip usually involves walking. Transportation planners have found that the walking part of the trip is normally perceived by the passenger to have a greater disutility than the part of the trip completed in a vehicle. Referring to a study in San Francisco, A. Ceder and J.N. Prashker of Israel state that people who were interviewed considered one minute of walking time to be 6.2 times less convenient than one minute of in-vehicle time. They further added that other studies found ratios of perceived disutility of walking to in-vehicle time ranging from 3.5 in Chicago to 2.0 in the Netherlands. The following related work is presented.

□ John P. Braaksma of Canada promotes the reduction of pedestrian walking time by outlining transportation tools to aid in the design of airport terminals.

□ Ceder and Prashker develop an algorithm to design the location of bus stops and terminals to minimise passenger walking distance.

□ Margaret C. Bell of England outlines the criteria to be applied in the design of interchanges at line-haul stations in suburban areas to reduce the perceived disutility by passengers during the pedestrian phase of the transfer.

□ Nils T.E. Rosén of Sweden emphasises the need to plan urban traffic facilities to favour the pedestrian phase of trips to a greater degree. And Eliahu Stern of Israel and J.J. Bakker of Canada indirectly allude to the same theme.

□ The paper submitted by L.R. Leembruggen describes the electric public transit vehicles proposed for the city of Brisbane in Australia. The author stresses the expected scarcity and increasing cost of fuel oil as an energy source for the motive power of vehicles as well as the increasing cost of providing fixed-rail transit systems as the reason for favouring this new type of battery-powered bus and combined battery-trolley bus.

□ The paper by Mr Louis Culer of Belgium on fire protection in metropolitan subways is, in my opinion, a manual on design principles and standards to guide the designer of subways in ensuring the minimum risk of fire or the consequences of fire in a subway.

□ Finally, the paper by W. Victor Sefton of Canada deals with the planning of parking garages in Toronto.

The last two papers are not reviewed further in this report.

SURVEY OF THE PAPERS

A. Ceder and J.N. Prasker. The object of this paper was to describe the development of an algorithm to yield the minimum number and locations of bus stations which would not require any passenger to exceed a pre-selected walking distance. Solutions to the problem of optimal stop locations have been suggested by several researchers, but the additional problem of a critical distance restraint was not part of their solutions.

The authors suggest that both problems can be solved by the same method and they detail the algorithm procedure mathematically step by step. In this problem, for which an algorithm is defined, there is a road network (arcs) in which the demand is generated in specific locations (nodes) along the arcs. The stations can be in the nodes or anywhere along the arcs. This is called an m-centre problem. The nodes serve as community locations and the stations of the public travel mode are located along the arcs. The authors conclude that further studies might consider a more realistic problem where the demand or passengers' origins are homogeneously distributed along the arcs rather than in specific nodes.

John P. Braaksma. The author states the case for participation by transportation engineers in the design of terminal buildings at airports. The airport consists of three transportation systems, the ground-side system, the air-side system and the system within the terminal building. The terminal building is the interface between the ground transportation system and the air transportation system. Besides providing shelter for the passengers, its transportation activity involves the transfer of passengers and baggage between modes. This transportation facility falls within the realm of the transportation engineer. This paper specifically concentrates on design tools which can be used by transportation engineers to assist in the design of terminals so that the function of passenger and baggage handling can be accommodated more efficiently. The author makes the point that there has been a dearth of quantitative tools to aid the designer in preparing the terminal for its traffic function. Having conducted, with others, research in this area for the past ten years, he summarises some of the more recent worn on airport terminal design tools. These tools range from heuristic layout methods to sophisticated simulation models.

He points out that the design task for a terminal building in respect of the traffic or transportation aspect falls within the realms of synthesis, analysis and evaluation.

Heuristic models have been developed for the design task of synthesis. For the analysis task analytical models have been developed, and for the evaluation task, simulation models. The author has developed a heuristic design tool to produce preliminary layouts of the air terminal building and contiguous apron. The design problem was to synthesise an optimum terminal concept in terms of minimum people-distance. A computerised methodology was developed of which the inputs are the expected daily flight schedule, a set of road factors, a connecting passenger flow matrix, a set of space standards, and a set of design and operating policies.

The problem was broken down into three sub-problems; facility sizing, flight assignment, and facility layout. The facility sizing problem was to find the minimum amount of space required, that is, minimising the capital cost of the terminal system. The flight assignment problem was to determine the best possible assignment of passengers and aircraft to facilities. Best was defined as least pedestrian transportation cost, i.e. minimising the user costs by minimising a weighted distance function. The facility layout problem was to produce a geometric configuration of the known facilities so that transportation costs would also be minimised. Algorithms were developed for each of these problems. The judicious assembly of these algorithms froms the design methodology. Having synthesised a layout, the next design step is to analyse this layout and the concepts contained therein. One of the tools used is the establishment of the trip-length frequency distribution of the pedestrian traffic flow pattern inside an air terminal. This flow is a function of the layout of the terminal and assignment of flights to facilities, such as gate positions and devices for claiming baggage.

The author next describes some models for analysing pedestrian traffic flow in the air terminal and some techniques for surveying pedestrian traffic. Another tool used is an analysis of curb operations. In airport terminal design one of the most difficult problems is to measure quantitatively the effectiveness of the curb in front of the terminal building. The author describes some of the approaches to this question and illustrates a method which has been used successfully.

The next stage of the design process is evaluation, for which simulation tools are required. The author points out that numerous computer simulation tools have been developed during the past ten years to help terminal designers evaluate their plans, and work is continuing in developing more accurate and sophisticated models. In his paper, the author indicates several of the simulation tools that have been developed and used effectively.

The author concludes by saying that after his brief overview of airport terminals, their design and design tools, the following observations can be made. First, airport terminals do not function in isolation. They are part of the overall transportation system and serve as intermodal transfer facilities. Secondly, since the primary function of an air terminal is to facilitate the transfer of passengers and baggage from one mode to another, much more attention must be paid to this function. Thirdly, form and function must be put in proper balance and transportation engineers can play an important role in the design of airport terminals. This is illustrated by the small sample of design tools which have been developed so far. Fourthly, the design process was considered to consist of three parts - synthesis, analysis and evaluation. Design tools suitable for each part were identified. There appears to be a need for further development of tools for the synthesis part of the design. Fifthly, although the emphasis of the paper is on engineering tools, it is recognised that the engineer is but one member of a terminal design team consisting of architects, human disciplines and engineers with expertise in civil, mechanical and electrical systems. All of these working together with proper tools can improve the quality of airport terminals.

Margaret C. Bell. The author develops some principal ideas and criteria for interchange design and planning. At the outset, the point is made that in trips from outlying suburban areas to high-density urban areas efficiency of travel by any one mode varies. For example, movement by private car is reasonably efficient in the suburban area but less so in urban areas. Similarly, bus travel over long distances in the suburban area can be inefficient owing to the need for many pick-ups and deliveries along the route. Again, in the inner city, bus travel can be inefficient in view of the conflict with congested traffic on the urban streets. The author states that in those circumstances a transit system with its own right-of-way through the congested area of the city is the most efficient. To derive the greatest benefits from the trip, the author suggests that the suburban trip should consist of travel to a station on the line-haul facility which is described as a train facility to take one into and through the city centre. An express bus with its own right-of-way is not ruled out.

The paper concerns itself with the interchange which takes place on the line-haul station at the point where the feeder meets the line-haul. This interchange is considered to be a change of mode from either private car or bus to the train. The paper explores various studies made by researchers on the elements of the trip at the station where the interchange takes place which are perceived by the user to be disadvantages. The purpose of the exploration is to determine what the transit user finds satisfactory or unsatisfactory, and the end object is to recommend criteria and designs to minimise the elements considered undesirable.

Five elements were considered: the walking trip between the point of disembarkation at the end of the feeder stage and embarkation in the line-haul station, the problem of queueing for information and tickets; the problem of scheduling the arrival and departure of buses and trains so as to minimise waiting times; the need for good access to the station area; and environmental factors within the station which would make it more attractive to the user. I should like to enlarge on some of the points made.

□ Access. The access to the station should be sufficient for both private motor vehicles and buses, and there must be an area for parking and kiss-and-ride facilities. This may require improvements to existing roads or the construction of new ones. It may require traffic control measures and careful layout of the location of parking areas, bus platforms and the kiss-and-ride facilities so as to minimise the walking distance.

□ Walking distance. The author suggests that studies made in other areas indicate that the walking distance should be restricted to no more than 250 m.

□ Queueing. In regard to delays caused by queueing for information and tickets, the author suggests measures such as through ticketing and adequate information signposting.

□ Scheduling. The author suggests skilful management of train and bus arrivals at platforms to minimise waiting times.

□ Environmental factors. The author suggests that these may be regarded as psychological instruments to induce the public to interchange by making a building attractive, with adequate heating and ventilation, by providing protection against the weather and by providing escalators, lifts and travelling sidewalks wherever they are possible. There should be services such as cafes, bookstores and waiting rooms. This all helps to reduce the perceived waiting time.

The author concludes that the interchange system of transport is an excellent way of providing a more flexible public transport system by exploiting the full potential of the existing transport network. Its concept is sound and, when implemented correctly and skilfully, it can give highly successful results, often without large capital investments. The paper develops the principal ideas and tries to formulate criteria for interchange design and planning. In the absence of research specifically to measure these criteria, they are derived empirically by measurements and observations made of successful interchanges. As such, they do not constitute actual limits but offer guidelines.

Nils T. Rosén. This paper departs from the general theme of design principles for transportation facilities in operation in that it deals primarily with planning concepts rather than matters relating to the physical aspects of transportation facilities and equipment. The paper is essentially critical of the present trend in urban transportation planning which, the author believes, gives the automobile pre-eminence in planning concepts. He strongly urges that the direction of planning be changed in order to place more emphasis on the bus and on cyclists and pedestrians. The author points out that it is necessary to take action against private downtown driving. He states that a downtown designed for buses is good for the unprotected traffic participants, while a downtown designed for cars can never accommodate a good bus system, or provide acceptable conditions for pedestrians, bike riders and residents.

The author promotes such measures as bike-ways designed specifically for the cyclist which will incorporate the shortest distance and exclusive rights-of-way, instead of being inserted in a road plan for private automobiles. Similarly, bus roads should be straight and well planned and exclusively for buses from one residential centre of gravity to another. As for geometric design at bus terminals and transit stops, the author points out that the bus passenger is transformed into a pedestrian and suggests that designs should be so oriented as to give pedestrians greater protection and direct access. The author claims that this change in planning direction can be accomplished incrementally over time without a violent upheaval in the present pattern of land use, but it is difficult to conceive of this, particularly in fully developed urban communities.

Eliahu Stern. The author describes attitude surveys undertaken in the Israeli northern Negev district, a sparsely populated area consisting of small urban centres and their adjacent rural hinterlands. In this area there is a low level of inter-urban bus service. The problem of servicing such an area is that the amount of traffic is so small that it cannot justify even subsidised transit services. A service must be provided, however, if only because of social norms and governmental commitments. The paper examines public preferences and satisfactions with regard to inter-urban bus services, and arrived at a proposed system which would satisfy the wishes of the patrons living in the area. Questionnaires were distributed to 753 citizens in both urban and rural settlements to obtain their response on five quantifiable service attributes: frequency, reliability, travel cost, connectivity and distance to nearest station.

The survey was conducted in 1977 and from it the author extracted preference and satisfaction profiles of service attributes. These profiles were based on 18 socio-economic homogeneous groups stratified by sex, age and income. The profiles were separated for urban and rural population and also by dominant mode of personal travel (captive transit and car users). The results indicated that the 18 preference profiles based on the socio-economic factors revealed no significant differences. Therefore, the profiles in this regard were aggregated. Significant differences were revealed between the profile for the urban population and that for the rural population, mainly in relation to the dominant mode of travel. They did not add significantly to the overall information, however, and a comparison of the profiles led to the conclusion that the profiles of the urban and rural population sufficiently represented the overall variations of preferences.

These preferences related negatively to satisfaction with the existing levels of service, which included items such as walking distance, frequency of service, reliability of service, time of travel and cost. For example, rural residents who now have to walk a distance of approximately 1.5 miles to a bus station showed a high preference for reduction of walking distances and were willing to rank the problem of connectivity or transferring throughout the trip much lower. People in urban areas with much shorter walking trips ranked the walking trip much lower in their preferences, and a reduction in the transferring required to complete their trip much higher. In addition, people in the rural settlements preferred a more frequent service, whereas the urban resident, who enjoys a reasonable frequency of service, ranked this factor much lower.

After examination and evaluation of the preferences of the population, a regional system of bus operation is proposed, and this system is then correlated with the preferences obtained in the attitude survey. The existing service is described as basically a radial system centred on the urban centres, with very little attention to inter-urban connections. The proposed system is basically an inter-urban main-line system with feeder buses - perhaps smaller vehicles - to penetrate the rural hinterlands to a greater degree. The author finally suggests that the proposed new regional transport system be introduced on an experimental basis, after which a second study should be undertaken to examine any possible discrepancies between attitudes and behaviour.

J.J. Bakker. The city of Edmonton had to plan for a high growth area in its north-eastern sector. The population in 1977 was 3,500 and was expected to increase to 50,000 by 1985. To provide a transportation facility for this sector, the options were a freeway which would include 70 buses in peak hours, including express services to serve the corridor; an all-bus service using 150 buses in peak hours, including express services through the central area of Edmonton; and a light rail transit system which would be integrated with the bus system on the basis of a no-cost transfer. The third option would require 75 buses in peak hours, mainly as feeders in cross-city service, and 14 LRT cars on the rail line.

In comparing the three options, the cost to the city was estimated, bearing in mind the various subsidies available to the city from the central government. In terms of this subsidy programme, the light rail transit with the integrated bus system proved to be the most economical as far as the city itself was concerned. Other considerations discussed were:

□ No change in free transfer system. One of the main principles in selling this system was the fact that a free transfer would be available between the bus system and the rail line or vice versa, thus providing one integrated fare system for the transit user. This system applied previously between the various bus and express routes within the city.

□ Minimum route changes. The major effort in the planning and design of the system was to minimise the disruption caused by re-allocating the routes of buses so as to service feeders to the LRT. In anticipation of the possiblity of a north-east LRT line, several major route changes were introduced in 1972 which, in effect, minimised the bus route changes required for the introduction of a light rail transit line. That is precisely what happened in 1978 with the introduction of the LRT. The routes in the residential areas could stay where they were but the express portions could be eliminated and diverted to the nearest LRT station.

□ Minimum transfer waiting time. A very important principle was to ensure that transfers with minimal delays in waiting time would be available to the user. The paper outlines how the schedules were carefully integrated in order to ensure this kind of minimal delay transfer situation.

□ Route selection followed the major desire line. The route of the LRT was selected along an existing railway right-of-way, since it provided the only reasonable location. There were several major traffic generators along the railway right-of-way, such as the stadium, the exhibition grounds and the Coliseum, but there were no generators in terms of residential or working locations. Nevertheless, the route does follow one of the major desire lines of travel for the north-east sector. The rail line intersects diagonally the north-south-east-west grid system of roads, and therefore provides a very satisfactory system of feeder buses to reach the various rail stations.

□ Public input. The paper describes the various inputs of the planning process, including the public input and its influence on the final systems design.

L.R. Leembruggen. The author describes an alternative to the current public transportation vehicles which is being proposed for the city of Brisbane in Australia. He discusses changes in energy availability and the greater sensitivity of people to environmental factors, such as pollution and noise, as well as the economic stresses which effectively reduce the potential of future expansion of the heavy rail public transit operations. These factors suggest that the street-car, the subway and, in particular, the diesel bus are no longer compatible and that an adaptable alternative is to be found in an electric vehicle known as the townobile. The paper cites the many advantages of this vehicle, compared with existing public transportation vehicles. The most important of these advantages is the fact that the vehicle does not require fuel oil for its power source, it is non-polluting and is more efficient than any other electrical vehicle in the sense that it has been designed specifically as an electrically powered vehicle.

The vehicle comes in three distinct configurations. One is a battery-operated bus, the second a trolley-bus and the third is a combination of trolley and battery operation. The combination of these vehicles permits great flexibility in their operation in that they need not be confined specifically to routes where trolley wires have been installed. Such trunk routes can be used by these vehicles which subsequently can leave these routes under battery power to complete the journey in areas where overhead structures might not be desirable for either economic or aesthetic reasons. The advantages of this vehicle also include easy servicing. In place of major service centres, it is proposed that a number of small curb-side facilities be provided where batteries may be recharged, tyres changed and other minor mechanical services provided.

The author concludes that, for the first time, there is an electric road passenger system which does not require expensive new permanent ways, extensive maintenance workshops and which, in its own right, can compete with noisy and polluting equipment in the areas of passenger attraction, cost of operation and flexibility of application.

CONCLUSIONS

A very few years ago a call for papers on the theme of "Design principles for transportation facilities and operations" would have elicited a large number of papers dealing with road traffic primarily involving the private motor vehicle. It is therefore significant that for this conference the theme almost exclusively elicited papers on public transportation modes of various types. The second observation I should make on the submissions is that in matters of facilities and operations, I would have expected to have received a number of papers from traffic engineers and transportation planners involved in operations. Surprisingly, the submissions on this theme were mostly from academics representing universities throughout the world.

With these two facts before us, it is important to analyse the significance of the papers which have been presented. Can we assume that the lack of submissions on private transportation indicates that all design principles for this mode have been fully explored and need not be given any further consideration at a conference of this stature? Or is it to be concluded that there is a swing away from concern for private modes of traffic to public modes? The second point to be analysed is the significance of the lack of submissions by practising engineers and the predominance of members of the university community among those who have submitted papers.

I believe that if anything has come out of this conference in respect of Theme 3, it is the question whether there has been a change in emphasis to public transportation, rather than the specific matters with which these papers were concerned. If indeed the prevalence of papers dealing with public transportation signals a change in direction and attitude the transportation engineers are taking or are expected to take, then this must be conveyed to all those participating in this conference. I would be prepared to accept such a conclusion if it were not for the predominance of academics among those submitting papers, for I must keep in mind that, in the event, it is practising traffic engineers who have the greatest influence on the direction in which traffic control, planning and regulation will be taking us.

I am also concerned by the fact that there were no papers by economists in relation to design principles for transportation facilities and operations. Surely, regardless of how desirable various transportation trends may be with regard to planning and perhaps even engineering principles, the economic impact of such plans is of great importance.

The papers presented for this theme were most interesting and illuminating in both the specific details to which they were addressed and the general concern expressed for the need to emphasise public transportation.

Operational measures to improve individual and public transport

23

B. Beukers Netherlands

SYNOPSIS

The paper offers a general report on the fifteen papers submitted under Theme 4 of the Tel Aviv conference, of which nine are included in the two volumes of the printed record.

INTRODUCTION

Only six of the papers were available for advance review and in so far as this survey refers to individual papers, only the contributions from Harvey Friedson, Fumihiko Kobayashi, Rudolf Lapierre, R.A. Reiss and others, J.A. Bourdrez and others, and Bosse Wallin will be mentioned.

GENERAL FRAMEWORK

Almost all the papers and abstracts submitted state that towns in the countries concerned are experiencing a deterioration of living conditions and a decrease in accessibility. Because of this, many town centres are losing some of their functions, while the older residential area close to the centres are exposed to serious decay. The measures being considered to check this disastrous development must at present be taken against the backdrop of a sharp fall in the growth of the urban population and bleaker economic prospects. This means that at present much less attention is being given to extension of the existing infrastructure and the introduction of new transport systems. The general approach is to make optimal use of the existing infrastructure and transport systems in existence.

Besides preserving the positive effects of traffic and transport on the economic, social and cultural life in the towns, we are also trying to reduce the negative effects considerably. Noise, air pollution, accidents, congestion and barrier effects are among the disadvantages. These negative effects have served to heighten public interest, as reflected in the increased participation by the people in decision-making processes concerning traffic and transport. Urban transport policies nowadays are management oriented and usually attempt to improve the urban environment by trying to discourage the use of private cars and promote the use of other modes of transport. This is also necessary because of the considerable rise in the cost of energy and other raw materials, while the social costs of noise, air pollution and congestion are clearly recognised. Besides this, the cost of providing and maintaining the infrastructure as well as running the public transport system is rising continually.

More and more towns have decided on a systematic approach. Proceeding from a number of formulated objectives, a comprehensive set of measures, instead of just one or a few types of measures, is drawn up and carried out. Improvement of the urban environment always ranks as one of the most important objectives. Nowadays this is achieved partly by excluding through traffic from residential areas and city centres. A very effective way to achieve this is to establish so-called environmental cells, or traffic cells, which result in through traffic being concentrated on a limited number of main roads. This, of course, intensifies the problems on and along these main moads; nor can any solution be found for the nuisance caused by parked vehicles.

A second objective, deriving from the first, is to cut down the use of private cars by imposing economic measures (such as parking fees) and by physical restrictions. A third objective is to encourage better utilisation of private cars and their infrastructure by car-pooling and staggering working hours, or the introduction of flexible working hours, which is more common in my country. A fourth objective is to ensure that traffic proceeds as smoothly as possible on the main highways by using area traffic control systems or corridor traffic control systems. A fifth objective is to encourage the use of public transport by improving the efficiency, regularity and frequency, the fare system and comfort. Reserved lanes and priority at crossings are essential to increase efficiency and regularity. A sixth objective is to stimulate the use of bicycles and mopeds by creating a dense network of cycle paths. Special attention should be paid here to special cycle routes, not necessarily sited along the main highways in order to avoid the noise and fumes of the traffic. A seventh and very important objective is the improvement of road safety by separating the various types of traffic as much as possible on the main arterial roads, by careful planning of crossroads and traffic regulations, and by reducing the speed of local traffic in residential streets.

It is clear that these objectives can be achieved to the desired extent only by the introduction of a package of complementary measures in which factors such as ease of application and the scope for enforcement should not be lost sight of, because none of these measures can have more than marginal effects if they are applied individually. To design such a package, it is also necessary for the objectives to be ranked by the political authorities concerned because some measures may complement one another while others may be conflicting. A package of complementary measures such as this must be carefully attuned to the specific circumstances of the town concerned, such as the location of activities, major shops and parking facilities and the street pattern. Not only does the application of these measures in the transport system of a town require careful planning, but it is also desirable to calculate the after-effects by before-and-after studies in order to be able to make any necessary corrections or adjustments. One thinks on one hand of direct effects, such as the impact on the quality of the urban environment and accidents, and, on the other, of indirect effects, such as retail sales and activity locations.

COMPREHENSIVE TRAFFIC MANAGEMENT PLANS
The papers by Bourdrez and others and by Wallin give examples of comprehensive traffic management plans in Sweden and the Netherlands. Both plans show how physical, economic and administrative measures are coordinated in order, on one hand, to accomplish the policy goals they embrace and, on the other, to ensure that all traffic essential for the effective functioning of the town and its economic life can continue.

The form and method of implementation of the plan for the city of Stockholm provide a very interesting study. The policy goals are to improve communications, reduce environmental disturbance caused by traffic, lessen the risk of accidents, facilitate public transport, and facilitate goods distribution.

In the present economic climate, measures which produce good results at low cost are given priority. The long-term plan for Stockholm envisages a fundamental reorganisation of the transport system. Through traffic will be restricted to a coarse network of arterials and major roads, while zones free of through traffic will be established. This zonal subdivision will not bring about any restrictions on public transport, cyclists and pedestrians. Heavy vehicles longer than 12 metres will be restricted to the major road network and streets in industrial areas. At night, between 10 p.m. and 6 a.m., vehicles with a total weight of more than 3.5 tons will also be restricted to the major road network and industrial streets. A network for cycle traffic has also been prepared. An astonishing fact, however, is that reserved bus lanes are open to cycle traffic. In my opinion this is only feasible if there are few buses and few cyclists.

The measures are to be implemented in stages. In the first stage, parking supervision and parking regulations are intended to reduce long-term parking. In addition, certain zones free of through traffic will be established and new bus lanes and priority for buses at traffic signals provided. New pedestrian facilities and cycle routes and paths will also be created. This first stage will take five to seven years to complete. In the second stage, supplementary measures will be initiated to limit traffic and promote the use of other traffic modes. These supplementary measures may include further parking regulations and restrictions or toll charges for cars entering the inner city. If these are still insufficient, even more drastic measures can be taken at a further stage. The Swedish paper concludes by giving some interesting views on the expected effects of the measures of each phase. The effects are classified in categories of road users, residents, industry and public bodies.

In the city of Groningen, in the Netherlands, the traffic management plan was implemented during one night in September 1977. In order to ban through traffic, the inner city was divided into four sectors for car traffic, while more pedestrian areas were created. Public transport, bicycles and pedestrians can still cross the sector boundaries. In addition, a new bus terminal for urban and regional buses was established near the inner city shopping area.

Several groups of the population, especially businessmen and shopkeepers, were very much against the traffic management plan because they expected that the adverse effects on the inner city would be considerably greater than the benefits. Partly for this reason, the Dutch government gave Groningen financial and moral support for the implementation of the plan and for an extensive before-and-after study of the quantitative and the qualitative effects. Attempts are being made to sound out the views of residents, workers and visitors on the changed conditions in the inner city, apart from determining such things as changes in traffic volume, modal split, numbers and types of visitors to the inner city, shopping habits, turnover, road safety and noise. The initial results of the study conducted after the plan was put into effect show that there has been no major change in the pattern of visits to the inner city, but that car traffic in the inner city has fallen to 60 per cent of the former level, although the number of parked cars in the inner city has remained about the same. However, the environment in the inner city has taken a considerable turn for the better.

CYCLING AS AN ALTERNATIVE MODE

In many countries the bicycle has been discovered or rediscovered as an attractive, healthy and cheap alternative to the car, especially for short distances. For this reason, cyclists are being given much attention in many traffic management plans, which often include the creation of a network of bicycle facilities. This network should be relatively dense because the cyclist is very sensitive to distances. If he has to make a detour, he may soon stop using the special provisions for cycles or even the bicycle itself.

Special provision for bicycles is necessary not only for the comfort of the cyclist but also for his safety. After all, the cyclist is, relatively speaking, very vulnerable, so that a confrontation with an automobile can often result in serious or even fatal injuries. Besides provision for cycling, adequate and strategically placed parking facilities should also be provided for bicycles - for example, near shopping areas, office buildings and schools. Cycle paths are traditionally laid out alongside main highways. In recent years, however, special cycle routes have been provided which are separated from the main arterial roads by one or more intervening rows of buildings in order to protect the cyclist from the noise and fumes of the traffic. Special cycle routes such as these have been built before in a number of new residential areas, but cycle paths between the residential area and the town centre are an innovation.

Friedson's paper gives a good example of the way in which a network of cycle facilities has been realised in a medium-sized town - Tempe, Arizona. As was the case with many other measures, adaptation of the plans and standards arose as a result of public participation in the decision-making process. The example of Tempe shows, however, that a good bike-way system can be created to provide access to schools, shopping centres and bike-way systems in other cities. Special mention should be made of the push-buttons that the city of Tempe provides for cyclists at all actuated signalised intersections.

TRAFFIC CONTROL

How to achieve the smoothest possible traffic flows on main arterial roads is the question currently on everyone's lips, especially in cases where the through traffic is to be concentrated as far as possible on a comparatively coarse network of arterial roads. This is very often one of the conditions for the success of a traffic management plan, since continuous congestion on the main arterial roads seriously affects the accessibility of towns and particularly the centres and, what is more, increases the environmental problems along these roads. It is not surprising, therefore, that three papers deal with electronically operated traffic control systems in some form or other. The conclusion drawn from these papers is that distributing the flow of traffic over the network plays an important part.

Lapierre's paper gives a summary of recent developments and experiences in Germany in influencing the traffic process, traffic routes and driving habits. The methods used are traffic signals, changeable direction signs and guide signs. Additional objectives are the optimal use of the existing road network; the improvement of traffic operations in terms of safety, economy and environmental nuisance; that priority be given to public transport and emergency services; the preference be given to the higher directional traffic flow during peak hours; the protection of certain traffic routes or areas, such as roads to and from schools and pedestrian precincts; the guidance of traffic towards a destination, for example a car park; and the screening of bottlenecks and congestion zones.

It is interesting to note that, apart from the traffic-dependent signal programme selection systems now increasingly in use in Germany, traffic-dependent signal programme generation systems are being tested. This is a new development in area traffic control. There is also a specific project to improve green-wave systems, i.e. synchronised traffic lights by an on-line optimising process, with a green wave generation model. The best results of the new types of control systems showed significant improvements in the mean travel time on a test section. The development of parking guide systems by which motorists are guided to open parking spaces via the shortest possible route is another significant innovation. Lapierre states further that, unlike the rerouting systems on the main highway network, rerouting on city streets has not worked out satisfactorily in Germany. He also describes the modal flow schemes and reversed public transport lanes in Germany.

Kobayashi's paper deals with part of a very comprehensive automobile traffic control project in Japan which was started in 1973. It singles out the route guidance system for urban networks. Some research on these systems has also been done in the US. The Japanese version, however, is very much more comprehensive. Thanks to the efforts of the Japanese government (MITRA), the whole set of software and hardware problems has been dealt with as a whole. The system is very modern, with a data-collecting facility able to retrieve the actual destination code from the car. Optimising strategies can therefore be continuously developed according to the actual traffic demand. It will be interesting to pinpoint the differences between this system and the cheaper systems which depend on historical statistical OD patterns. It is important to note that the intention is not primarily to benefit the individual but to alleviate congestion when rerouting traffic.

A great deal of work had to be done on a computer model that works out the best routes. The model works with two groups of traffic: first, the groups of vehicles which cannot communicate with the system and whose OD trips have to be calculated on assumptions about the drivers' wishes, and, secondly, the vehicles which have given their destination code and which must continue to receive directions in good time. The studies show how important the right choice of an optimum rerouting strategy is. It is very interesting to see the results of the simulation studies on the effectiveness and efficiency of a dynamic routing system, especially in its relation to the route selection algorithm.

The American paper by Reiss and others gives a very good description of the development of an algorithm for corridor traffic control, in which the corridor facilities can consist of motorways, motorway frontage roads and signalised arterials. Because of the complexity of the overall process of traffic control in such an inter-city corridor, a hierarchical structure has been chosen, which consists of two levels of control.

Corridor level controls optimise traffic flow on a large scale by distributing the traffic among the various corridor facilities, and local level controls optimise the use of individual facilities independently of one another, but always based on the predicted demand as determined at the corridor level. The corridor level control consists of three distinct modules, namely demand estimation, corridor control and corridor control message selection. The local level control consists of four distinct modules, namely ramp metering control, arterial signalisation control, incident control and state estimation. The corridor level control algorithm dynamically assigns traffic to corridor routes in order to optimise a selected performance criterion or objective. It also generates a diversion policy.

An important point is that the algorithm has been set up in such a way that the results become more accurate when more time is left for the calculations. This means that at any time the best possible rerouting suggestions are available, which is very important for on-line control. This paper is a good example of thorough analysis of the problems, followed by careful evaluation of the solutions and strategies with the assistance of simulation techniques before they are applied in practice.

The changing role of the transportation engineer

24

John E. Baerwald   United States

SYNOPSIS

The paper offers a general report on the seven papers submitted under Theme 5 of the Tel Aviv conference, four of which are included in the two volumes of the printed matter.

INTRODUCTION

The seven papers submitted under this theme are by Hans B. Barbe, António José de Castilho, Alessandro Orlandi, Horst Sternberg, Richard J. Brown (Paper 4), Boris Dobrer and Norman Abend. The first five were available for advance review. The papers tend to complement one other while discussing different viewpoints on the theme subject. While their emphasis is different, the basic opinions form a remarkably integrated whole.

SURVEY OF THE PAPERS

Hans B. Barbe. The traffic engineering profession, a newcomer when compared to medicine and law which are over 2000 years old, has gone through four transitions since its inception in the early 1920s. The first transition is identified as that of traffic engineer to traffic planner, with the term traffic engineer referring to the common public association of this function with traffic operations (such as increasing capacity and safety through traffic control devices and regulations and intersection channelisation.) In the late 1950s, with the concurrent development of land-use planning, the correlation between land use and trip generation caused the traffic engineer to become involved in the development of comprehensive traffic plans.

Soon, it became apparent, however, that something was missing in these plans and the next transition was under way. This transition, from traffic planning to transportation planning, flowed from the change from developing basically highway-oriented traffic plans to multi-modal transportation plans. These changes occurred earlier in Europe because public transport was a much more effective transport mode in European cities than it was in US urban areas. The development and analysis of alternative multi-modal plans was facilitated by the concurrent rapid maturation of electronic data-processing equipment and techniques.

Increased recognition of the very important role played by transportation in the total social system led to the third transition, from transportation planning to integral planning. The rapid growth of urban areas led to the need to consider more than the coordination of land-use planning and multi-modal transportation planning, Economic, political, social and other constraints must also be part of the plan development and evaluation process. The use of multi-disciplinary study teams increased, and greatly increased opportunities for public participation were introduced.

More recently, recognition of the worldwide problem of environmental pollution led to the fourth transition, from integral planning to environmental planning. Concern for air, noise and visual pollution, preservation of historic sites, and many other factors led to environmental impact evaluations. Requirements for energy conservation must also be considered. Thus, the transitions from the primary concern for better traffic operations to environmental planning have taken place in a relatively short period of fifty to sixty years.

António José de Castilho. Growing awareness of the increased complexities of modern living in Portugal, as in other countries, has led to an evaluation of the changing role of the transportation engineer and how he can become better qualified through formal and continuing education.

There are still insufficient linkages and coordination between the various interest areas within transportation technology, including public and private modal operating agencies, consultants, contractors, suppliers, researchers and educators. There are also deficiencies in coordination between administrators, planners and designers, operators and other functional units within an agency or company. Further compounding the problem is the misapplication of foreign technology. There is also the question of how specialised and how generalised the education of a transportation engineer should be. Some examples of cooperation between varied transportation interests and specialities in studies under way at the Labóratorio Nacional de Engenharia Civil are described.

In training courses for transportation engineering students and professionals, social as well as technical and economic considerations in problem-solving efforts should be included. This is critical because of the growing complexity and influence of environmental, energy, safety and other human problems on transportation engineering and planning.

Alessandro Orlandi. Although there have been continued efforts to improve analytical, modelling and evaluation procedures, the logic scheme for territorial (regional) transportation planning is still based on that developed in the late 1950s during the Chicago and Detroit regional transportation studies.

It is now necessary that transportation be considered a single phenomenon and that we consider and analyse all aspects of transportation in a single, disciplined fashion. A generalised model has been developed, which contains three logical phases: the scope, or the type of study to be done; the field of interest, which represents the sector in which one must operate; and the basic components, which represent the overall knowledge available to perform the job. The successful transportation function must be accomplished while satisfying the conditions of safety and regularity.

Transportation is a cause-and-effect relationship. As such, it is generated as a consequence of a desire to carry out a transfer in the desired manner and time. Thus, one can formulate the following postulate on the general principle of transportation motives: generation of the transfer demand; and effecting the transportation and fulfilling the demand. From this postulate one can derive the methodology to be used when studying any transportation problem: to know the requirement for movement and the circumstantial conditions in which it is generated and by means of which it develops; and to study and create the system which will accomplish this transfer within the stated conditions.

The entire mechanism of generation is expressed in a generalised model for the study of transportation problems. In applying the model, extreme care should be exercised in predicting future conditions such as population size and movement desires and the status and characteristics of existing or future transportation systems. In studying transportation problems, the transportation engineer should function as a transportation expert who considers all the diverse aspects which any transportation problem may have, rather than strictly as a planner, organiser or designer.

Horst Sternberg. Solving traffic tasks in West Germany is almost exclusively done by construction or civil engineers. Recent changes in the academic programmes of universities, academies, and technical colleges permit traffic-oriented degree candidates to specialise in either road construction or the municipal building trade. Student who wish to become officials of boards of road works at the federal or state levels will major in road construction. Students interested in urban traffic will major in the municipal building trade option, with an opportunity for intensive study in two subject areas, town streets and city railways, which include town and transportation planning courses. Thus town planning is no longer the exclusive domain of architects, but may also be done by traffic engineers. This will allow interdisciplinary development of town plans and also provide for more public participation in the planning process.

The traffic engineer of today is not only faced with increased public interest in his work, but he must also work with many different public and private agencies in order to reconcile his efforts with concurrent and often conflicting environmental, social and other considerations. The development of alternative solutions and a basis for evaluating these solutions has to be done on a multidisciplinary basis which includes traffic engineering participation.

It is also important that all urban traffic engineering functions and duties be combined into a single agency headed by the chief traffic engineer. Traffic planning and town planning agencies must be equal and cooperative participants in the development and implementation of comprehensive plans and programmes.

Richard J. Brown (Paper 4). The urban transportation planning process has recently been criticised on the grounds that after considerable expenditure on major studies, there has been little or no improvement in travelling conditions in the study areas. The major problem area in the transportation planning process is not the details of the analytical methods employed, but rather the organisational structure within which the process is managed. Among the organisational problems encountered are fragmentation of the various professional groups involved in the planning process, the futile introduction of political dogma into transport thinking, the confrontation between the broad requirements for travel of a metropolitan area and the needs of the local communities within it, and the lack of co-ordination between the various operators. The goal should be a system wherein the management of the planning process will ensure the correct balance between the politicians as the executive, the professionals as the analysts, the community as the consumers, and the operators as the managers.

Certain key functions have been identified as components of a logical decision-making process. They are assessing, planning, organising, implementing, evaluating and revising. These decision-making functions involve four phases: gathering information, drawing conclusions, making decisions, and execution. Thus, those who will be involved in the execution stage should be made aware of the issues requiring a decision and encouraged to participate in the data gathering in accordance with their particular expertise. Decision making should be decisive and systematic and should place strong emphasis on a diagnosis of the whole situation. The gathering of information and the drawing of conclusions phases are concerned with the planning function. The effectiveness of this function must be assessed in the following areas: time, perspective, involvement, data collection and analysis, and objective setting.

There are essentially four easily identified participants in the transport sub-system, namely the politician or executive, the public, the planner, and the operator. Their relationships through six two-way links and four activity planes are described by the author. Each of these planes represents a clearly identifiable function in the management process. Of prime importance in the organisation structure of any public body is a high degree of political accountability. Depending on the situation, this may be accomplished at local, regional, state or national levels. The public representatives and thus the public of a metropolitan area should be wholly involved in the decisions affecting that area.

The author recommends an organisational structure to meet the transport tasks of a metropolitan authority. In many areas at present the expertise or manpower is not available to set up this type of organisation, but immediate efforts should be initiated to step up the resources available for the education and training of managers within the various transport-oriented disciplines. A better balance should also be sought between the public and private sectors in the transport fields so that the private sector is used as a source of specialist knowledge rather than a source of manpower.

COMMENTS AND CONCLUSIONS

The first officially recognised traffic engineers appeared on the American scene in the early 1920s. They were usually part of state or city road departments and their primary concern was traffic operations. In some cities the traffic engineering function was located in the police department and often it is still positioned here, especially in communities of less than 100,000 which do not have professional engineers in city employment.

Recognition of the need to correlate roadway design with vehicle and driver operating capabilities led to the traffic engineer's involvement in geometric highway design. Similarly, the traffic engineer's participation in transportation and community planning activities increased as knowledge was gained about the relationships between trip generation and land use. With increased recognition of the need for multi-modal considerations in providing the best transportation service, the transportation engineer came on the scene.

Whenever new technical specialities develop from older disciplines, the transition period is often marked by conflicts and misunderstandings because there is a tendency on the part of those in authority to continue in the old ways and routines while the new concepts and techniques are being developed. Jurisdictional disputes and inter-departmental jealousies occur. The advocates of the new technology must prove themselves competent and better able to serve the public needs. This prolonged process is still being experienced by traffic and transportation engineers in several countries.

As we consider the transportation and other major problems now facing mankind - social justice, housing, population growth, pollution and others - we must recognise that there is an increasing tendency for citizens to question those in authority and to seek more self-determination for the people. As traffic and transportation engineers, we are now called on to practise our profession under the growing scrutiny of a public which is resisting increased taxation while simultaneously demanding more and greater public services. I therefore believe that in future the traffic and transportation professional will have to be aware of and concerned with the total effects of his work, not merely the technical results. We have no choice if we are to maintain, much less expand, the position of public trust that we so earnestly desire. We must not merely tolerate but actively encourage citizen participation in the development of transportation planning and operational programmes.

If we as traffic and transportation engineers and specialists are to play an increasing role in transportation decision making, we will have to make more effective use of our opportunities to participate in the total process of government. Generally, the initial opportunity is in the area of problem definition, i.e. the area of data collection, analysis, interpretation and projection. Unfortunately, after these activities have been completed, we often tend to retire from the public scene. Continued involvement will be enhanced if we are prepared to discuss objectively the implications of our technical desisions for the total framework of the community and the desires of society. We should be prepared to evaluate the potentially negative as well as the positive results of our own work. In order to do this, we have to be able to make similar evaluations of other projects which are also competing for the limited public funds available.

There are numerous examples of men who are now in key decision-making positions at all levels of government whose background is in traffic and transportation planning, design and operations. I predict that there will be an increasing demand for administrators with this background because society as a whole will come to recognise the increasing need for effective transportation as a means of obtaining a better life for all. For example, in the past many state highway departments in the United States named as district and central office administrators those engineers who had strong backgrounds in the fields of construction and maintenance because these fields were the major functions of the department. Now there is increasing emphasis on the planning, design and operation of these transportation facilities. Hence, we find more and more engineers with traffic and transportation engineering backgrounds being named to these decision-making positions.

As people movers, I envisage traffic engineers working more closely with men in other disciplines and serving as public advocates. I see us playing a leading role in interpreting human needs and desires to other professions, such as translating human requirements and capabilities to the civil or highway engineer who builds and maintains the roadway, and the automotive, mechanical and electrical engineers, who design and build the vehicles. I see us working much more closely with regional planners and architects as we all strive to create a much more livable environment. In order to fill the role of public advocates, we will have to become much more familiar with the physical, emotional and intellectual capacities and desires of the human beings we serve. We must develop much closer relationships with such man-related sciences as psychology, sociology and medicine. We also must become much more sensitive to the total needs and desires of society.

As traffic engineers gain increased opportunities to advise governmental decision-making bodies, we must be sure that our recommendations are timely, appropriate, practical, legal, understandable and realistic. As previously mentioned, we must be willing to discuss the negative as well as the positive aspects of our recommendations, and we have to be able to evalute our programmes in relation to other social needs and priorities.

As we begin to understand better the complexities of many of the problems facing mankind, we become more and more impressed with the fact that the only solution to most of these problems lies in interdisciplinary cooperation. It is therefore necessary that we as transportation and traffic engineers become much more conversant with the functions, interests, and terminology of such diverse but allied professions as urban and regional planning, architecture, land development, law, political science, education, enforcement, psychology, sociology, medicine, and systems analysis, as well as related engineering fields. I foresee the interdisciplinary team concept being expanded to include many other transportation activities beyond transportation planning.

Are there better ways in which we can anticipate future living requirements and travel desires? How can we motivate improved driver and pedestrian behaviour? What strategies and techniques will prove successful in encouraging the selection of a transportation mode which might be of greater public benefit at some personal sacrifice? Obviously, the help of many varied professional disciplines is needed by the traffic engineering profession to answer these and similar questions.

Expanded application of the team concept cannot be legislated for nor can it be a paper organisation. Its success will depend on the full and free participation of competent persons from all the professions best able to contribute to the solution of the problem. It must be a working relationship. In each of our contacts with other professional disciplines and with the public, we must be willing to listen and learn, for true cooperation can only be built on a foundation of mutual understanding.

We must also expand our efforts to recruit the most alert and aggressive young minds to the traffic and transportation engineering profession. We must inform these students of the enormous challenges and opportunities which exist. We must actively support responsible university officials in developing sound formal and continuing educational programmes, and we have to provide additional sources of financial assistance for qualified students.

The traffic engineer is no longer a glorified policeman. He is concerned with securing the most efficient and best use of existing land transportation facilities and he is equally concerned with the proper planning and design of new transportation systems and their interfaces. We must continue to improve our professional competence so that the public, through their designated leaders, will actively encourage traffic and transportation engineers to assume an increasing responsibility in governmental and private decision making. Merely talking to ourselves about our ability to serve is not sufficient, for it is by our active involvement in current affairs and the results of our work that our worthiness will be evaluated.

To summarise the various discussions on Theme 5 - The changing role of the transportation engineer - we can conclude that there have been dynamic changes in the role of transportation and traffic engineers during the short period of their professional existence. This changing role has not been recognised to the full by many of our employers, clients or colleagues, or by the general public. Thus, the credibility of our profession is at stake and we must live up to our professional responsibilities. We must be responsible for and recognise the importance of our decisions. We must continue to improve the quality and quantity of transportation engineers. Academic programmes must be encouraged, supported and monitored by our professional organisations and by each individual in professional practice. We must expand the opportunities for continuing education through conferences, meetings, short courses, graduate study programmes and other organised learning experiences.

In all our efforts toward transportation problem-solving we must consider broad as well as detailed elements, while honestly seeking to identify and quantify negative as well as positive results of the various alternative solutions. We should continue to develop more effective and responsive organisational arrangements through which transportation and traffic engineers can make their greatest contribution to the betterment of society. It is essential that we actively work toward greater public involvement in the transportation decision-making process. If this involvement is to be effective, we must use all practical means to keep the public informed of our activities, problems and desires. We cannot be true and responsible public advocates if we try to remain isolated and unapproachable.

The role of transportation and traffic engineers is indeed changing. These changes are not only occurring in the conference subject area of urban planning but they are also happening in the allied interest areas of design, operation and management. How well and how rapidly these changes take place will directly depend on how responsive we are as a profession and, most important, on each of us as individuals. In either case, it is our challenge and our responsibility. We must and we shall prove equal to the task.